AF795703

Cognitive Technologies

Managing Editors: D. M. Gabbay J. Siekmann

Editorial Board: A. Bundy J. G. Carbonell
M. Pinkal H. Uszkoreit M. Veloso W. Wahlster
M. J. Wooldridge

For further volumes:
http://www.springer.com/series/5216

Antonio Krüger • Tsvi Kuflik
Editors

Ubiquitous Display Environments

Springer

Editors
Antonio Krüger
German Research Center for
Artificial Intelligence (DFKI)
Saarbrücken
Germany

Tsvi Kuflik
University of Haifa
Haifa
Israel

Managing Editors
Prof. Dov M. Gabbay
Augustus De Morgan Professor of Logic
Department of Computer Science
King's College London
Strand, London, UK

Prof. Dr. Jörg Siekmann
Forschungsbereich Deduktions- und
Multiagentensysteme, DFKI
Saarbrücken, Germany

Cognitive Technologies ISSN 1611-2482
ISBN 978-3-642-27662-0 ISBN 978-3-642-27663-7 (eBook)
DOI 10.1007/978-3-642-27663-7
Springer Heidelberg New York Dordrecht London

Library of Congress Control Number: 2012940253

© Springer-Verlag Berlin Heidelberg 2012
This work is subject to copyright. All rights are reserved by the Publisher, whether the whole or part of the material is concerned, specifically the rights of translation, reprinting, reuse of illustrations, recitation, broadcasting, reproduction on microfilms or in any other physical way, and transmission or information storage and retrieval, electronic adaptation, computer software, or by similar or dissimilar methodology now known or hereafter developed. Exempted from this legal reservation are brief excerpts in connection with reviews or scholarly analysis or material supplied specifically for the purpose of being entered and executed on a computer system, for exclusive use by the purchaser of the work. Duplication of this publication or parts thereof is permitted only under the provisions of the Copyright Law of the Publisher's location, in its current version, and permission for use must always be obtained from Springer. Permissions for use may be obtained through RightsLink at the Copyright Clearance Center. Violations are liable to prosecution under the respective Copyright Law.
The use of general descriptive names, registered names, trademarks, service marks, etc. in this publication does not imply, even in the absence of a specific statement, that such names are exempt from the relevant protective laws and regulations and therefore free for general use.
While the advice and information in this book are believed to be true and accurate at the date of publication, neither the authors nor the editors nor the publisher can accept any legal responsibility for any errors or omissions that may be made. The publisher makes no warranty, express or implied, with respect to the material contained herein.

Printed on acid-free paper

Springer is part of Springer Science+Business Media (www.springer.com)

Preface

Nowadays, we increasingly live in smart environments – computerized environments that can sense, track, and model their users and, as a result, provide them with personalized services. Computers have become small enough to be embedded in everyday objects, from shirt buttons to pencils; high-quality (and soon to be flexible) computer display devices can be embedded in objects of all sizes, from wristwatches to billboards; and wireless Internet communication is becoming more widely available every day. We are also transitioning from traditional linear media to highly targeted, interactive media. The convergence of these factors (miniaturization, display technology, wireless communication, and interactive media) will lead to environments where media and visual content will not only be accessible on desktop computers but at many different locations in a given environment. Such ubiquitous display environments will support a rich variety of interactive display devices in order to provide their users with relevant information in a seamless manner. This will enable people to interact with information artifacts, rather than with dedicated information-processing devices, in a more natural and casual fashion.

In this book, leading researchers in ubiquitous display environments envision a day when a wide variety of displays will seamlessly provide carefully targeted information to users. For example, as you enter a building, a display shows you a carefully selected list of events, in which you may be interested. When exiting the building, a display reminds you where you parked your car and recommends the best route home. This convergence of technologies poses exciting and fundamental research challenges. It will lead to innovative technologies for human education, cultural heritage appreciation, and scientific development.

This book is based on talks given at the German–Israeli Minerva School in autumn 2009 at the University of Haifa, Israel. It addresses the challenges of how to exploit these new technologies within the context of an educational and cultural experience. It discusses both the scientific and the technological aspects of these challenges. On the scientific side, it integrates Artificial Intelligence, User Modeling, Temporal and Spatial Reasoning, Intelligent User Interfaces, and User-Centric Design methodologies. On the technological side, it integrates mobile and wireless networking infrastructures, interfaces, group displays, and context-driven adaptive presentations.

The book covers a diverse set of topics, as the issue is multidisciplinary in nature – from multiagent systems and architectures to aesthetics, from reasoning about time and space and about visitors' behavior to actual design, implementation, and evaluation of context-aware information delivery over large displays. While it was impossible to cover all aspects of ubiquitous display environments, we have tried to sample the state of the art in the field and we hope that it introduces some additional aspects.

May 2011

Tsvi Kuflik and Antonio Krüger
Haifa and Saarbrücken

Contents

Ubiquitous Display Environments: An Overview

Tsvi Kuflik

Abstract Ubiquitous display environments are public spaces with various situated public displays. These displays are intended to provide relevant information to people in their vicinity, where these may include the regular inhabitants of the space as well as visitors. Their nature makes the development of such displays a multi-disciplinary task that combines technological with aesthetic, sociological, and legal aspects. Research usually focuses on specific aspects while neglecting others, since it is intended to explore only a limited part. Here we try to take a step back and portray the big picture, showing how various aspects need to be addressed and integrated in order to support us in today's and tomorrow's ubiquitous display environments.

1 Introduction

We are beginning to live in smart environments – computerized environments that can sense, track, and model their users and use these models to provide personalized services. This is mainly because computers today have become small and cheap enough to be embedded in everyday items from shirt buttons to pencils. Moreover, high-quality (and soon to be flexible) computer display devices can be embedded in objects of all sizes from wristwatches to billboards. Wireless Internet communication is becoming more widely available every day. We are also transitioning from traditional linear media to highly targeted, interactive media. The convergence of these factors (miniaturization, display technology, wireless communication, and interactive media) will allow us to leave our desktop computers behind and make the transition to a radically new computing paradigm – the *ubiquitous display environment*, as envisioned by Weiser [1]. These smart environments will support

T. Kuflik (✉)
The University of Haifa, Mount Carmel, Haifa 31905, Israel
e-mail: tsvikak@is.haifa.ac.il

A. Krüger and T. Kuflik (eds.), *Ubiquitous Display Environments*,
Cognitive Technologies, DOI 10.1007/978-3-642-27663-7_1,
© Springer-Verlag Berlin Heidelberg 2012

a rich variety of interactive display devices in order to provide their users with relevant information in a seamless manner.

This is one of the most exciting and important areas of technology for human development. It will enable people to interact with information artifacts in a more natural and casual way as compared to dedicated information processing devices. For example, as you enter a building a display may show you a carefully selected list of events in which you may be interested. When exiting the building a display may remind you where you parked and recommend the best route home.

This revolutionary convergence of technologies presents exciting and fundamental multidisciplinary research challenges. It will lead to new, innovative technologies for human education, cultural heritage appreciation, and scientific development. This book presents a few aspects of this multifaceted research challenge.

2 Overview

In Weiser's seminal paper [1] introducing ubiquitous computing he describes how "The most profound technologies are those that disappear. They weave themselves into the fabric of everyday life until they are indistinguishable from it ... Specialized elements of hardware and software, connected by wires, radio waves and infrared, will be so ubiquitous that no one will notice their presence." This means that computers and communication infrastructure are becoming an integral, invisible, and unobtrusive part of the environment. When this process is complete, the focus of computerized technology will change from the means (computers) to the services provided to the users – new ways of delivering information and the natural interaction of the user with their environment. Many different areas of research need to be combined in order to make progress here. Some are in fact areas that have been researched for a long time while others are new; some involve deep theoretical background while others involve exploratory studies about how users may use and interact with novel technologies. By its nature, research on ubiquitous display environments, where users interact naturally with information sources in public spaces, is a design study: ideas are formed, prototypes developed and tested experimentally, and the results used for further improvement.

2.1 Context Awareness: Time and Space

Dey and Abowd [2] claim that "While most people tacitly understand what context is, they find it hard to elucidate". They defined context as "any information that can be used to characterize the situation of an entity. An entity is a person, place, or object that is considered relevant to the interaction between a user and an application, including the user and applications themselves." Among the most important contextual aspects, issues that involve reasoning and acting within the time constraints of the real world are the most challenging. Temporal information may be qualitative, where events are represented by abstract time points and time intervals, and we process and deduce "relationships" between them, such as pairs intersecting each other, one preceding, following or containing another, etc. Other information may be quantitative, where durations can be measured, precise time stamps may be available, or numerical methods can be applied to understand a specific time line of events. These temporal aspects are discussed in chapter "Perspectives on Reasoning About Time". The challenges and lessons learned from developing a context-aware ubiquitous display application are discussed further in chapters "Context-Sensitive Display Environments" and "xioScreen: Experiences Gained from Building a Series of Prototypes of Interactive Public Displays" of this book. Chapter "Context-Sensitive Display Environments", reflects on the development and presents different display environments that were used as research platforms to investigate various aspects, including technical issues and also the question of how to improve interaction with display environments and their usability. Chapter "xioScreen: Experiences Gained from Building a Series of Prototypes of Interactive Public Displays", studies the interplay between shared public displays and private handheld devices, and explores how such technological settings would be used by students in daily life in a university campus environment when accessing context-aware information services.

2.2 *Individual and Group Interaction with Large Displays*

Interaction of users with large displays situated in public spaces is multifaceted and as such requires multidisciplinary research. On the one hand, in order to be effective, the communication should be personalized, but obtaining information about a first time (and maybe also last time) mobile user in a ubiquitous computing environment is challenging. This is because personalization relies on the availability of information about the individual (or group). Ubiquitous user modeling research tries to address this challenge. Chapter "Challenges and Solutions of Ubiquitous User Modeling", provides a brief introduction to user modeling and its techniques and then elaborates on ubiquitous user modeling, its challenges, and the state-of-the-art research. It aims to provide a broad introduction to key approaches to user modeling and personalization as a starting point for the reader who wants to appreciate the challenges of personalization for ubiquitous display environments. However, in public spaces there are more than just individuals. Chapter "Audience Measurement for Digital Signage: Exploring the Audience's Perspective", reports on the user's perspective of the audience measurement techniques employed for digital signage, where two digital signage systems were deployed, 1 for a year, and the other for 5 years. Chapter "Analysis and Prediction of Museum Visitors' Behavioral Pattern Types", reports on applying unsupervised machine learning techniques for analyzing the behavior of visitors to cultural heritage sites, in order to explore ways of generating an effective user model quickly without any prior knowledge in a non-intrusive manner by using predefined stereotypes. Chapter "The Design, Deployment and Evaluation of Situated Display-Based Systems to Support Coordination and Community", reports on experiences gained from two situated display-based deployments and explores some of the issues that arise when such systems are used on a day-to-day basis. It presents insights about the social nature of public and semipublic spaces both before and after the introduction of situated display technology, the interactions that manifest themselves (over time) in the domain and the long term use of novel technologies, especially their collaborative and community effects. While the above aspects are natural when discussing the technology, there are additional issues that are not less important. Technology should be trusted by its users and this is the focus of Chapter "Trust Management of Ubiquitous Multi-Display Environments". It presents a decision-theoretic approach to trust management that is considered particularly appropriate when a system has to balance the benefits and risks of a decision carefully. The chapter presents decision policies that help maintain trust in critical situations, and presents their implementation in three interactive applications that have been developed as part of a university-wide ubiquitous displays management system. While trust is important, chapter "Considering the Aesthetics of Ubiquitous Displays", suggests that aesthetics is important as well, and its importance stems from three main perspectives. The design perspective suggests that beauty is a cornerstone of any design discipline and that it is often positively related to other practices of good HCI design, such as usability. From a psychological perspective it argues that aesthetics induces pleasure and satisfies human needs, and that it is influential in how people form first impressions and subsequent attitudes. Finally, it argues that aesthetics has an important practical role as a differentiating factor between interactive products.

As we can see, interaction with large displays is indeed multifaceted and we touched upon a few aspects, while others, such as privacy, need to be addressed as well.

2.3 *Infrastructure: Agents*

Ubiquitous display environments, as any ubiquitous environment, are highly dynamic. Users tend to enter and leave the environments at will, in many cases for the first and last time in their lives, services appear and disappear, devices and systems may be available or not, all of which requires a highly dynamic and flexible infrastructure. Multiagent systems seem to be a possible solution to the required dynamics and flexibility, as suggested by Busetta et al. [3] and many others. However, such agents need to interact not only with each other but also with humans and this is the focus of chapter "From Research to Practice: Automated Negotiations". While the development of proficient automated agents has flourished in recent years, making the agents interact with people has still received little attention. This is mainly due to the unpredictable nature of people and their negotiation behavior, though complexity and costs attached to experimentation with people, starting from the design and ending with the evaluation process, are also a factor. Even so, succeeding in designing proficient automated agents remains an important objective. In recent years, much effort has been invested in facilitating the design and evaluation of automated agents interacting with people, making them more accessible to researchers. The chapter describes the creation and evaluation of two distinct environments for bargaining agents, while proposing a novel approach for evaluating agents.

2.4 Unusual Visionary Application

When discussing situated displays, we usually think of large displays, installed in a train station or an airport, providing information to users passing by. However, this technology has many more potential uses and chapters "Shared Interfaces for Co-located Interaction" and "Virtual Technologies and Empowerment of Users of Rehabilitation" present two different usages. Chapter "Shared Interfaces for Co-located Interaction", focuses on the use of large displays, table tops in this case, for group interaction and specifically for co-located collaborative work. It introduces the difference between multi-touch and multi-user interfaces and presents three cases of co-located interactions on shared interfaces. The first two cases present examples of multi-user applications: StoryTable is a multi-user application using which children can practise their narration skills in a collaborative way; the *Narration and Negotiation Reconciliation Table* is a multi-user application used in peace-education programs designed to foster a shift in the attitudes of participants via a narration task. A new approach to shared interfaces explicitly aimed at influencing immediate behavior in an informal, non-goal-oriented co-located small group is then presented. Conversely, chapter "Virtual Technologies and Empowerment of Users of Rehabilitation", presents a completely different use of large displays: to empower people with disabilities. The goal of empowered rehabilitation is to guide users to achieve their maximal independence by developing the skills necessary to overcome the physical, social, and emotional barriers in their lives. In this chapter, a model of patterns of social change has been applied to show how the use of novel information and communication technologies may contribute to social change and the empowerment of people with disabilities. Each of the model's four patterns – social planning, social reform, social development, and social action – is illustrated with an example of the role of technology during the process. Lessons learned during the process of empowering people with disabilities may also be incorporated into general information and communication technologies programs to help enhance usage in additional segments of the population, including minority groups, as well as those who have traditionally been challenged in this area (e.g., the elderly, people who live in remote locations).

3 Summary

Ubiquitous display environments are a multifaceted and challenging domain. In today's ubiquitous computing environments, they may be used for a variety of different tasks. However, they are situated in dynamic and changing environments, where they have to provide personalized and context-aware services to their users; hence they should be carefully designed, and be supported by an appropriate infrastructure. This book aims at presenting the complex nature and diversity aspects that are involved in the development of such environments and serves as a reference for those interested in such environments.

References

1. Weiser M.: 1991, The computer for the 21st century, Scientific American, 3(265):66–75.
2. Dey A.K., Abowd G.D. Towards a Better Understanding of Context and Context-Awareness. Proceedings of the 1st International Symposium on Handheld and Ubiquitous Computing, Karlsruhe, 27–29 September 1999, pp. 304–307
3. Busetta P., Kuflik T., Merzi M., Rossi S. Service Delivery in Smart Environments by Implicit Organizations. The First Annual International Conference on Mobile and Ubiquitous Systems: Networking and Services (MobiQuitous 2004), August 22–26, 2004, Boston, MA, USA, pp. 356–363.

Challenges and Solutions of Ubiquitous User Modeling

Tsvi Kuflik, Judy Kay, and Bob Kummerfeld

Abstract Ubiquitous display environments are public places with various situated public displays. These displays are intended to provide relevant information to people in their vicinity, where this may include the regular inhabitants of the space as well as visitors. For this latter group, it is particularly challenging to provide relevant information. This is because personalization relies on the availability of information about the individual (or group). Ubiquitous user modeling research tries to address this challenge. This chapter provides a brief introduction to user modeling and user modeling techniques and then elaborates on ubiquitous user modeling, its challenges, and the state-of-the-art research. It aims to provide a broad introduction to key approaches to user modeling and personalization as a starting point for the reader who wants to appreciate the challenges of personalization for ubiquitous display environments.

1 Introduction

There are many drivers for personalization, but a dominant one is to help people cope with information overload. This problem has been steadily growing with the wide availability of computers and Internet connectivity. In addition, we are now starting to live in "active environments" [60], where the environment senses and responds to us – its inhabitants. This recent development, often called pervasive or ubiquitous computing, offers the promise of new ways to deliver information to people within their current environment, be it their home, office or a public space.

T. Kuflik (✉)
The University of Haifa, Mount Carmel, Haifa 31905, Israel
e-mail: tsvikak@is.haifa.ac.il

J. Kay • B. Kummerfeld
University of Sydney, NSW, Australia
e-mail: judy.kay@sydney.edu.au; bob.kummerfeld@sydney.edu.au

A. Krüger and T. Kuflik (eds.), *Ubiquitous Display Environments*,
Cognitive Technologies, DOI 10.1007/978-3-642-27663-7_2,
© Springer-Verlag Berlin Heidelberg 2012

Pervasive computing further aims to personalize the information made available to match the individual's current context and task, as well as their individual preferences and taking account of their knowledge, goals and needs. Emerging large displays that are embedded in the environment could be used to deliver timely information to the people in that environment.

Providing personalized information on ubiquitous displays is challenging. This is because personalized applications within a pervasive computing environment must acquire relevant information about the people in the space, so that they can use that information to drive the personalization. When such information is stored explicitly, rather than embedded within the logic of a program, it is called a *user model*. In general a user model may represent various forms of information about the user, such as their current needs as well as their long term preferences, knowledge and goals.

In the case where new users come to an active environment, the applications within that environment will have no prior knowledge of the users. Acquiring user modeling information is quite difficult. Certainly, it would be unacceptable to subject the users to a questionnaire to bootstrap their user model. Instead, a new framework for user modeling is required – "a ubiquitous user modeling framework" which can both capture relevant information about the user in the active environment and make relevant parts of a user model available when and where it is needed. This has been described by Heckmann [34] as follows: "ubiquitous user modeling describes ongoing modeling and exploration of user behavior with a variety of systems that share their user models".

Recent research in user modeling recognized this need and work has started in exploring approaches and techniques for supporting users in active environments. There are three main challenges posed by such active environments:

1. Little initial information about the users that request/need a service;
2. The heterogeneity of existing user modeling information;
3. The need to keep track and adapt a user model continuously for every user in the environment.

Moreover, taking a broader view, it is clear that there are critical challenges associated with effective management of privacy. This is tightly coupled with personalization, since the user model holds personal information which is subject to legal requirements. Privacy is also inextricably linked to security as well.

2 A Short Historical Perspective

User modeling research can be dated to the mid-1970s [49], when technological progress made personalization feasible and research began to explore how to provide personalized services to their users beyond the "one size fits all" paradigm. Importantly, one of the key foundations for that work was based on the recognition of the value of representing information about the user in an *explicit* user model [51], as well as a separate user modeling system to manage the model.

Such a user model could hold arbitrary personal information, such as the user's age, education, income, life style, interests, plans, preferences, past interactions with the system, and any additional relevant information needed by the system to provide a personalized service.

A key task is the initialization of a user model in such systems. Accurate initialization of a user model is important to every system that provides personalized services. In some cases, it is important to reuse the user model. For example, if the user rarely interacts with an application, it needs a means of acquiring the user modeling information efficiently, without bothering the user. In such systems, stored user modeling data from previous interactions is usually not available at the outset. We now focus on ways that user modeling research has developed to address this problem.

2.1 Basic User Modeling Representation and Reasoning Techniques

In more than 30 years of research in user modeling, a variety of user modeling techniques have emerged. Some of these provide ways to represent the user model. Generally speaking, representation and reasoning in user modeling are tightly coupled; hence in every section we will briefly discuss the relevant representation issues. We focus mainly on a selection of techniques that support efficient *reasoning* about the user, especially in cases we start with limited information about them and need to infer more to support the personalization. We have chosen these to illustrate the character of the approaches available, rather than provide an exhaustive review. These approaches include "content-based", "case-based", "collaborative" (or "social"), "demographic", "knowledge-based" reasoning, and various hybridizations of them, as described below, following [31]. These differ in their complexity, the amount of information they keep and their pros and cons as presented below. In addition the use of activation/inhabitation networks and overlay models, that are common in eLearning and cultural heritage will be discussed.

2.2 Feature-Based and Content-Based User Modeling

In a feature based approach, a user model is a set of feature-value pairs, representing various aspects of the user, such as interest in specific sport, music type, level of knowledge in a specific area, and more [9]. This can be at different levels of detail, as needed by the application that matches values of features representing items/information suggested to the user with the features in the user model, in order to determine their similarity or relevance. One commonly used approach for measuring the similarity of items and users assumes the users and items are represented as vectors in an n-dimensional space and the cosine between these two vectors represents the similarity or relevance. This is used to rank the items to be suggested/recommended to users. The content-based approach of user modeling as described by Hanani et al. [31], is a specific variant of the feature based approach. It takes into account the user's areas of interest, as represented by the textual content of documents the user has previously indicated are relevant or interesting. From this, it creates a richer user model as a basis for personalizing services – e.g. the user model is a set of terms, possibly weighted, representing the area of interest. This approach was initially described, as the "cognitive approach" by Malone et al. [59]. In general, content-based systems analyze the textual content of documents that users rated, to infer the users' informational needs. These systems deduce their recommendations by matching the analysis of user needs with the analysis of available rated items content.

2.3 Case-Based User Modeling

Agnar and Plaza [3] described the general case-based approach as utilizing the specific knowledge of previously experienced, concrete problem situations (cases) where these has been "captured" in a way that enables the system to recognize similar cases later. Humans use this approach as a powerful and frequently applied reasoning mechanism.

The case-based approach covers a wide variety of techniques for organizing, retrieving, utilizing, and indexing knowledge retained from past cases. Cases may be kept as concrete experiences, or as a set of similar cases that form a generalized case. A solution, based on a previous case, may be directly applied to the current case, or modified according to the differences between past and present cases. Matching and adaptation of cases may utilize a deep model of general domain knowledge, shallow and compiled domain knowledge, or no domain knowledge at all. A technique may be purely automatic or depend on interactions with the user.

For a user model, a case is associated with personal characteristics of the user involved. An example may be a tour plan that contains, in addition to the tour plan itself, some personal characteristics of the users who planned the tour. The system can then match new users to previously stored cases to recommend the same plan to other users with similar characteristics, as in the case of Trip@dvice[1] [68].

2.4 Collaborative User Modeling (Based on Matching Similar Users)

The underlying assumption of the collaborative approach is that users with mutual taste who agreed in the past tend to continue to agree in the future [27]. This is

[1] http://www.ectrlsolutions.com/web/guest/tripadviceportal.

a relatively simple user modeling approach that is widely used in deployed recommender systems (for example, Amazon – people who bought this item also bought…). According to this approach, the user model is a list of ratings for items – a vector of ratings, where the dimensions are the individual items (e.g. books, movies, etc.). Following this approach, a system compares a user's model to other users' models by matching items they liked or disliked in order to identify similar users. Once similar users are identified, the system attempts to predict how much the current user will like an item that has yet to be rated. The prediction is accomplished by computing a weighted average over all the item's ratings given by similar users. This approach somewhat resembles the case-based approach. However, it differs from the case-based approach since instead of finding the most similar cases it gives an average score for each item of interest using every available case, without considering its similarity. (The focus is on the similarity of the behavior of the users; hence it is sometime called social filtering [31].) This means that users with correlated preferences are sought. The collaborative approach is useful in domains that contain a variation of interests, such as movies and books. Systems using this approach categorize users with similar preferences as similar users, and use past interactions of similar users to provide recommendations. The collaborative approach suffers from the *cold start problem*: users have to rate several items before they can benefit from the system and items need to be rated before they can be recommended.

It is worth noting that in addition to the *user to user* matching discussed above there is also an *item to item* approach, where the user's ratings for similar items are averaged for prediction of the rating of the target item (instead of averaging ratings of similar users, see Ref. [71]).

2.5 Demographic User Modeling

The demographic approach relies on a marketing approach which postulates that users with similar demographic backgrounds may have similar preferences. This approach uses an analysis of demographic data about users, who rated a certain item, to learn and categorize the "type" of a person who likes the item. This information is stored and then used to provide for future recommendations. Although demographic data is used for marketing, its use in user modeling is relatively limited, since it is *stereotypical* by nature and requires a reference demographic data base that is usually not available. However, in some cases it has been used in research to enhance recommendations in hybrid user model based systems (see below).

2.6 Knowledge-Based User Modeling

Knowledge-based systems infer their recommendations from a knowledge base about products and typical users [13]. Systems applying the knowledge-based approach utilize a domain-specific knowledge base and analytical skills of human experts to advise users who are novices in the field (in performing tasks/searching for information etc.). These systems require domain knowledge that is a result of a knowledge engineering process, which is known to require significant effort from an expert and is time consuming. A domain expert is required to choose a method of representing the knowledge, a method of utilizing it, and a method of structuring it [12]. These methods should allow the system to mimic the experts' solution to a problem. In general, knowledge based systems were quite a focus of considerable research as expert systems in the past but the problems of eliciting, representing and updating such systems means that they have less importance than the other approaches described where automated analysis of data provides models that can be readily updated.

2.7 Hybrid User Modeling

The hybrid approach strives to exploit the benefits of complementing approaches, and simultaneously aims to avoid their downfalls, as surveyed by Burke [14] in the case of recommender approaches. Systems using the collaborative approach need to ramp up with sufficient amounts of data, otherwise they are susceptible to erroneous performance. Until a sufficient number of ratings are collected, the system might not be useful for users for whom there is insufficient data from collaborating peers. A knowledge base generated by an expert may be used to avoid the collaborative approaches "ramp up" problem. A drawback of knowledge-based systems, which the collaborative approach facilitates solving, is the indifference of knowledge-based systems to individual preferences. This could result in a knowledge-based system suggesting a common solution, even when the user exhibits an uncommon behavior. The combination of these approaches facilitates better personalization than would be provided by one of these approaches alone. The drawback of this type of hybrid system is that knowledge engineering is still required as is the case of the pure knowledge-based approach.

2.8 Activation/Inhibition Networks

This approach, when applied to personalization, is tightly coupled with a rich domain knowledge base. A user model is overlaid on a domain ontology, that may be structured, for example as a taxonomy, semantic net or other knowledge base. When there is a description of a domain, represented by a tree or a graph of linked concepts ("domain ontology"), an individual's user model can be represented as a partial graph/tree with the concepts that the user is modeled as knowing or being interested in. The representation may model each node of this model in various ways, for example, a simple Boolean or a numerical value. The user model is adapted by a reasoning mechanism that reasons and propagates new evidence about the user preferences/knowledge level (based on positive/negative feedback) between concepts (preferences, knowledge) along the links from one concept to all the concepts linked to it, applying heuristic decay mechanisms (so propagation stops after a few steps). Overlay user models have been particularly common where systems need a detailed mapping of users' knowledge or interests, in areas like eLearning or cultural heritage.

2.9 Stereotypes and User Modeling

One important notion in user modeling, identified and named by Elaine Rich [69] is the stereotype. This captures the notion that a small amount of information about a user can be used to infer a large amount of other information. For example, if we are told that a person is a judge, we can infer they are highly educated, well to do, honest, and the like. This basic idea was used to create a double stereotype for knowledge of the Unix operating system [91] where observing that a person used sophisticated commands was used to infer they were an expert and knowing a person was an expert was used to infer a default stereotype representing a rich knowledge model, indicating many things that user would know. Similarly, if a person indicated they were novices, this single piece of knowledge was used to infer a detailed default model of their knowledge. This was used to support a natural language advisor and it is representative of an important part of the early user modeling research which was motivated by the demands of natural language dialog. It is also representative of work on intelligent tutoring systems, an important body of work that has spanned the whole of user modeling's history [73]. The notion of stereotype was more tightly defined [44] as a set of default assumptions that are made when a small set of triggers becomes true, with the definition of retraction conditions which can be used to determine that a stereotype should no longer be used. The key idea was that a stereotype should be used to address the cold start problem, by using a small amount of information about the user to infer a rich, statistically valid, initial model that would be used only until more reliable information about the user became available. It was noted that such a stereotype could be used to reason across domains, for example, taking a model of the user's preferences for books and using this to infer a model for the their preferences for movies.

3 Ubiquitous User Modeling and Its Challenges

3.1 Introduction to Ubiquitous User Modeling

While user modeling as a research area began about 35 years ago, the ideas of ubiquitous computing (or pervasive computing or ambient intelligence) have been around from the late 1980s [88]. During the 1990s, topical workshops and conferences on ubiquitous/pervasive computing appeared, followed by new journals and magazines such as IEEE Pervasive Computing that started at 2001.

Ubiquitous user modeling has closely followed the evolution of ubiquitous computing. There was recognition of the link between the goals of ubiquitous computing and the need to support users in such environments by taking into account personal aspects as well as contextual ones [41].

One of the important common ideas in ubiquitous computing and user modeling is that a useful model of a person may well need to draw on evidence from a range of sources. This was central to the um toolkit [43, 45], which distinguished evidence in terms of the nature of its source, for example treating stereotype-based evidence as less reliable than observations of the user's actions, which in turn might be treated as less reliable than information the user explicitly gave to the system. Similarly, Vassileva [84] used stereotypic user models that integrate task and expertise characteristics to bootstrap a user model that was later updated based on user behavior. van Elst et al. [83] suggested augmenting business processes that usually rely on role and task information in order to support their users, with personal information. Hence to integrate user modeling data from two sources – stereotypical contextual information, related to the task and role, with information about the specific worker, for better service. These ideas were used for pervasive computing to model location [18], making use of multiple sources of varyingly reliable sources if evidence, including Bluetooth and system sensors, both prone to various forms of unreliability but together being useful to infer the user's location.

In order to provide a system that can accept evidence about aspects of the user, one needs a framework. The first steps towards this were user modeling shells [49], which provide a representation for the user model and various user modeling tools that can be reused by multiple applications. It should be understood that at the time of that work, applications tended to be stand alone and specific, and user modeling capabilities were integrated into the application. The first step into ubiquitous user modeling was made by decoupling the linkage of the application and the user modeling component and introducing the general user modeling shell systems [49].

During the late 1990s, commercial personalized systems started to appear, with some of these having a client-server architecture. This might be seen as an initial step towards sharing and re-using user data for personalization by different applications [7, 49]. In his survey, Kobsa [49] notes the need for mechanisms to import and export existing user data as a requirement from user modeling server, but without suggesting any mechanism or framework for that process. He also stated correctly that processing done by personalized systems at that time cannot be used outside the context of the specific domain and application due to the lack of abstract representation of learned users' characteristics.

Kobsa [49] detailed requirements that will facilitate wide dissemination of generic user models. (Although he distinguished academic and commercial work, the requirements are complimentary and we integrate them. Technical performance requirements were not included).

- Generality: domain independence, compatibility with as many applications and domains as possible, and for as many user modeling tasks as possible.
- Expressiveness: ability to express as many types of facts and rules about the user as possible.
- Inferential capabilities: capability of performing various types of reasoning and resolving the conflicts when contradictory facts or rules are detected.
- Import of external data: ability to integrate the user modeling data collected by the system with the data collected by other systems.
- Privacy: support of privacy policies and conventions, national and international privacy legislations, and privacy-supporting tools and service providers.
- Quick adaptation: ability to quickly adapt services to new users, personalization functionalities, applications, and domains.
- Extensibility: Provide Application Programmer Interfaces (APIs) and interfaces that allow for the (possibly bi-directional) exchange of user information between user-modeling tools (allowing the integration of variety of user modeling techniques).

Kobsa [49] concluded his survey of generic user modeling systems with fairly accurate predictions of the development of networked computers and especially mobile computing. He suggested two options for ubiquitous user modeling with a user model residing on the server side or even on the mobile device. Furthermore he introduced the notion of personalization of smart appliances, and the potential of multiple-purpose usage of users characteristics and discussed in light of this the pros and cons of client side versus server side user models [93].

Kobsa [49] concludes: "...one can expect to find a wide variety of generic user modeling systems, each of which is going to support only a few of the very different future manifestations of personalization and other applications of information about the user..." The obvious conclusion from the above is that given the expected variety of limited user modeling servers on the one hand, and the usefulness of re-using available precious user modeling data on the other hand, there will be a need for some kind of generic mechanism for user modeling data sharing, conversion and exchange. This is because we cannot expect that all systems and servers will use the same user modeling ontology/language. We refer to such mechanisms as user modeling mediators. Such mechanisms are the essential foundations for ubiquitous user modeling since the very basic need for ubiquitous user modeling is to accumulate and exchange user modeling data with variety of applications, as needed for ad-hoc personalized service delivery.

3.2 Challenges of Ubiquitous Computing

This section provides a very brief overview of ubiquitous computing challenges, in order to introduce the challenges of ubiquitous user modeling. In Weiser's vision for ubiquitous computing, computing artifacts will disappear "They weave themselves into the fabric of everyday life until they are indistinguishable from it." Weiser [89] presented various infrastructure challenges for mobile computing, such as power problems, wireless communication, means of interaction and privacy. All of these have had considerable work since that time. Infrastructure issues are being tackled in various ways (for example, [23, 30, 94]), solutions also have been suggested for ubiquitous collaboration [40], privacy in ubiquitous computing is an important and ongoing area [15, 56, 57], design and evaluation of ubiquitous services [39], integration of context awareness into pervasive computing applications [63], development framework for small screen mobile devices [66], framework for interaction and visualization across small and large screen devices in ubiquitous computing [75] and even pervasive games [81] and more.

In parallel with the various solutions being proposed, new models, visions and challenges continue to appear [1, 35, 55, 70, 72, 87, 89]. There have been papers in which the "state of the art in ubiquitous computing" is reviewed and future research directions are identified [2, 64].

By its nature, ubiquitous computing aims to support individuals in computerized environments. As such, knowledge about the users, their needs, background, preferences, knowledge, etc., is necessary for personalized services. Hence for providing services to users in ubiquitous computing, novel technological solutions, including context awareness, are not enough and classical user modeling aspects need to be addressed as well [41].

3.3 *Challenges of Ubiquitous User Modeling*

Ubiquitous user modeling must overcome three major barriers – user modeling issues related to *user modeling techniques differences, domain differences* and *contextual differences*, as well as the need to efficiently and effectively bootstrap a user model for ad-hoc services and continuous updating of the user model. Let us use a brief example to illustrate the above requirements. A user is searching for a music CD as a birthday present for a friend. The user enters a store that they have not visited before. The store has a collaborative recommender system. Consider the first time the user accesses this system. Suppose the user has previously used a content-based music recommender system and his/her content based profile is available for the collaborative music recommender system. To make this useful for the store system, there is a need to convert the existing content-based user model to a collaborative one – or "cross technique user modeling mediation[2]". Now assume that instead of the content-based music user model, there is a collaborative movies user model (e.g. taken from movielens[3]). Now the challenge is how to

[2] Mediation is explained in detail in the next section.

[3] http://movielens.umn.edu/.

convert the movies user model to a Music user model – "cross domain user model mediation". The third aspect can be illustrated as follows, assume that there is a user model for that user at that online store – the user purchased several CDs for personal use in the past. However, this time they are searching for a present for a friend – a different context that probably requires a different user model (the friend's user model?), hence this is "cross context user model mediation".

It is worth noting that even systems that utilize the same approach may use different techniques in applying them. For example, two content-based systems may differ in their weighing scheme. One system may weigh every term extracted from the available contents and the other may categorize the terms, weighing the categories instead. Or, two book recommender systems may rate the books on two different scale – one on a scale of 1–10 and the other from −2 to +2. Consequently, different systems may well store user modeling data according to their own representation.

In an active environment, the envisioned scenarios may be a mix of representation, domain and context (e.g. to recommend a restaurant for dinner with a friend, using the collaborative approach, a system may need to integrate content-based information about books with restaurant preferences for lunch with colleagues). Hence, the available information may be from different domains, in different contexts, and represented in different techniques due to natural heterogeneity of the systems and services available and the dynamic contexts of the users.

Assuming that there are user models for this user, in various domains and applications, the challenge of ubiquitous user modeling is how to efficiently and effectively make use of these to construct an ad-hoc user model to drive a personalized service for a first time user in an active environment, from various heterogeneous user modeling sources?

Research in ubiquitous user modeling is trying to address this question in various ways, as described below.

4 Bridging the Gap: State of the Art of Ubiquitous User Modeling

The user modeling community recognizes the need for data sharing, based on standards and/or mediation. This is needed to enable and improve personalized services by making user modeling data available to a variety of applications beyond a single application which defines its own user model. Even though semantic web ideas offer the promise of ease of data sharing and possible standardization, techniques are needed to bridge the gap between the ideal semantically unified user modeling domain and today's practice – heterogeneity of user models, as recognized in Kobsa's requirements for user model data import and extensibility [49]. Beyond that, for supporting users in active environments when personalized services are delivered as needed, accurate user modeling data is essential. Among current research projects we can find attempts to standardize user modeling terminology in order to allow easy collaboration by the use of common ontologies and languages, general servers that allow partial views, as needed, to applications, importing and exporting user modeling data, hybrid user models, adding contextual aspects to better support users by taking into account their specific context and user modeling mediation. The following sections will survey the state of research.

4.1 Common Semantics for User Modeling

Common language is a key issue in integrating information from different sources in every domain, including, of course, user modeling. Ontologies, common languages and communication protocols are among the commonly accepted approaches to achieve this, while the advent of the semantic web provided a common platform that encourages and supports them. The state-of-the art approach to the problem of standardization of domain specific knowledge representation is the use of ontologies. Ontology-based representation of user modeling was discussed in Ref. [46], which motivated ontology-based reusable and understandable modeling of students. Reusability makes use of the separation of the user model from the personalization application or application domain. The structure of the user models was based on a set of predefined ontologies that facilitated access to a customized explanation of the meaning of the user modeling components in each domain. However, in spite of the great potential in the use of ontologies, they have not become widely used in user modeling tasks, possibly due to the initial effort required in the construction of domain ontology.

Dolog and Nejdl [21, 22] proposed an approach where a learner model can be distributed and can reflect features taken from several standards for a learner modeling. These features can be combined according to the requirements of specific personalization techniques, which can be provided as personalization services in a peer-to-peer learning network. They analyzed various usage scenarios and defined conceptual schema for learner feature categories. Denaux et al. [20] pointed out in a position paper the potential of combining ontologies for student modeling, especially when they follow standards.

The evolution of the semantic web provided the needed infrastructure for information standardization and sharing, including user modeling data. Henze and Kriesell [37], Henze and Krause [38] proposed a service-based architecture for bringing methods and techniques from the area of adaptive hypermedia to the Semantic Web (thus extending the adaptive hypermedia framework to the adaptive web [10]). In their framework, personalization functions from adaptive hypermedia were made available as web services, which a user can subscribe/un-subscribe to as they prefer. They have implemented their ideas in a Personal Reader, a framework for defining rule-based personalization algorithms for Semantic Web applications.

Their framework includes a Dublin-core[4] or LOM[5] based domain ontology and attribute-value pair user modeling ontology, an observation ontology, and an adaptation ontology. In their demonstration they provide a learner device while relying on LOM, user profile information is relying on the IEEE PAPI[6] specification for describing learners as building blocks and on RDF[7] for information sharing, thus standardizing the access to the user modeling data. This makes it potentially available to other applications following the same standards. However, the standardization of user model relies on IEEE PAPI hence it is limited to the tutoring systems domain and cannot be generalized beyond that, unless another standard (such as GUMO) is adopted.

The notion of ontology-based user models was further developed by Razmerita et al. [67], who presented a generic ontology-based user modeling architecture called OntobUM. OntobUM integrated three ontologies: a user ontology characterizing the users, a domain ontology defining the relationships between the personalization applications, and a log ontology defining the semantics of user-application interaction. A similar approach for ontology-based representation of the user models was presented by Heckmann [34]. He introduced GUMO, a comprehensive set of General User Model Ontologies, which allowed uniform interpretation of distributed user models in intelligent environments. GUMO represented user modeling data using the RDF[8] based OWL,[9] and was used for multiple personalization applications which operate at the same time. Such commonly accepted ontologies simplify the exchange of user modeling data between personalized applications and overcoming the inherent problems of syntactical and structural differences between their user modeling representations. The diversity in the area of user modeling also triggered the User Modeling Meta Data Ontology initiative UMMO [96] that was meant to structure the state-of-the-art in the field and serve as a central reference point and as a tool to index systems, papers and learning media. Such an ontology is beneficial for both the user modeling research community and the students, as it creates a shared conceptualization of the known approaches to building user models and their implementations (it seems that it was abandoned, since its links no longer function). A similar approach was also taken by several other researchers: Weiβenberg et al. [90] suggested the use of a complex hierarchy of ontologies for personal recommendation of events to mobile users; Mehta et al. [61] suggested using a common ontology for standardization of user modeling across systems; Brusilovsky et al. [11] suggested use of a common

[4] Dublin Core, 2004. http://dublincore.org/.

[5] LOM: Draft Standard for Learning Object Metadata, 2002. http://ltsc.ieee.org/wg12/index.html.

[6] IEEE P1484.2 Learner Model Working Group, PAPI Learner, draft 7 specification, 2003. http://ltsc.ieee.org/wg2/papi learner 07 main.pdf.

[7] Resource Description Framework (RDF) Schema Speci_cation 1.0, 2002. http: //www.w3.org/TR/rdf-schema.

[8] Resource Description Framework, available online at http://www.w3c.org/RDF/.

[9] OWL http://www.w3.org/2004/OWL/.

ontology for representing users' domain knowledge to be used by different personalized eLearning systems; Hatala and Wakkary [32] used an ontology for tagging domain objects and to allow reasoning about the user that is more abstract than reasoning based on content; Zhang et al. [97] suggested using web services as a mechanism for exchanging user modeling data among systems sharing a common user modeling ontology; Carmagnola et al. [17] used a common ontology, so allowing users to tag objects and share information for socially-based personalization. All the above (and many more) involve an initial intensive effort in creating the relevant ontologies. An alternative approach to manual construction of ontology is to learn a light-weight ontology, by mining available materials in the pervasive computing context [65].

The need for standardization of user modeling was recognized beyond the user modeling research community. As early as 1999, the need for standardization of user data for possible exchange of customers' data, lead to standardization work by an industrial consortium. The goal of the work was to support vendor-neutral interchange of customers' data for e-businesses and at the same time to provide a framework enabling privacy safeguards. The result was the definition of Customer Profile Exchange (CPExchange) specification for the privacy enabled global exchange of customer profile information. The proposed standard uses the Extensible Markup Language (XML) to format messages transmitted under standard Internet protocols and includes both a Customer Information model and a Privacy Information model. The information models contained in this specification facilitate customer profile transport and include the metadata that associates data protection (privacy) with customer profiles. The specification builds on the W3C XML Schema and the W3C P3P specifications [8].

As demonstrated above, there is an abundance of work exploring ways to apply social and semantic web techniques for user modeling interoperability (for further reading see a recent survey by Torre [80]). However, all face the need for agreement upon user modeling and domain ontologies; at this point, such agreement has not been achieved.

4.2 User Modeling Servers

The progress towards user model servers can be seen to have started with work on "user modeling shells", with their generic components for user modeling [48]. However, once it was realized that the personal information is important as well, user models that store personal data and provide it to applications for personalization appeared.

In 1995, Kay developed the um toolkit [45], a mechanism for reusing generic user modeling data for different applications. The core of the system is a repository of information about the user and this was stored in the user's own filespace. Each application can interpret the user modeling data in different ways using a "resolver" function that interprets the evidence about each component in the model. There were several built-in resolvers but the application could also apply its own.

This tool evolved into Personis [47] and its descendants, where the key difference is that a server stored a collection of user models for many users. Notably for pervasive computing, PersonisAD allowed for models that could be distributed [4]. Authorized applications could request access to the model. If that access was allowed, multiple applications could use and reuse the allowed parts of the model. The shift to the server approach created the need for additional support for privacy management. This was achieved by the use of filters that determined which classes of evidence were available to which applications and which resolvers could be used by them. If the user allowed general access to an application, it could have all evidence and make its own interpretation of that evidence. On the other hand, a very restricted application may only be allowed access via one resolver.

Fink and Kobsa [24] discuss the benefits of user modeling servers that have:

> Up-to-date user information for holistic personalization; Synergistic effects with respect to acquisition and usage of user information; Low redundancy with respect to application and domain independent information; Low redundancy with respect to stereotypes and user group models; Increased security; Increased support for the holistic design, acquisition, and maintenance of user models.

As discussed earlier, Kobsa [49] surveyed characteristics of academic and commercial user modeling shells and identified requirements for future servers for reuse of user models such as is needed for mobile and pervasive settings.

Work in user modeling servers started in the mid-1990s as generic systems and during the 2000s changed to deal with provision of user modeling data to variety of applications that need personal information, as can be seen in the work of Kay et al. [47], Fink and Kobsa [25], Carmichael et al. [18], Kobsa and Fink [50] and others. However, many issues remained unresolved, including the need for applications to adapt to the server's user modeling representation. This seems to be a major constraint since there is no active server that systems can use and so recommender systems and other personalized service providers produce their own user modeling mechanisms.

4.3 User Modeling Mediation, Interoperability and Hybridization

The ability to integrate fragments of existing user models is a possible solution for bootstrapping a user model when there is no prior personal information, but partial information may be requested from other systems. Berkovsky et al. [6] coined the term "user model mediation". The basic idea is that in many cases fragments of user models are available, in different representations for the same domain, or from different domains or even from different contexts, hence there is a need to define methods for translating or mediating them across representations, domains and contexts. They suggested and demonstrated several such techniques for transferring data from a collaborative to a content based approach and from a case-based to a content-based approach and discussed the more complicated situations (cross-domain and cross-context), suggesting future research in addressing these issues.

Carmagnola and Dimitrova [16] and Walsh and Wade [86] discuss user modeling interoperability – an approach that differs from mediation in the way it integrates user modeling data. Carmagnola and Dimitrova [16] suggest that "evidence" of user modeling data is collected from various sources and then a domain ontology is used to reason and integrate the evidence for creating an ad-hoc user model. Walsh and Wade, on the other hand, suggest an approach that allows automatic translation of user modeling data, following initial mapping between different models/representations/systems. In such a scenario, an administrator performs an initial mapping and then, as user models are updated and information is needed, up-to-date information may become available for any application that needs it, based on the initial mapping. Walsh and Wade suggested it for learners' models, where once mapped, ongoing relatively long time interaction can benefit from the initial manual mapping.

Vassileva et al. [85] suggest an approach of match makers – broker agents that are able to find the relevant user modeling information needed for a personalization task among a group of specialized user modeling agent.

Hybridization of user modeling techniques has less to do with user modeling interoperability and reuse and more with combining the benefits, and avoiding the limitations, of individual techniques. Notably, using complimentary techniques may overcome the cold start problem. This is relevant to ubiquitous user modeling – when there is a need to construct, from scratch, an ad-hoc user model. Hence, hybrid approaches deserve mention in this context as well (this approach was already introduced earlier, and for detailed discussion see Ref. [13]).

4.4 Using Social Networks Data for Ubiquitous User Modeling

Social networks offer an opportunity for systems providing personalized service to their users to gain publicly available personal information for bootstrapping a user model. This is extremely important in the ubiquitous computing scenario. Using the API of social networks, systems can request personal information and by applying a suitable mediation mechanism, convert it to their domain and internal representation. This was explored for several different social networking sites with particular care to ensure users can understand the system [53, 54]. There is considerable amount of work exploring the potential of social networks data for user modeling. Liu et al. [58] were among the first who tried to extract personal information available in social networks and apply it to personalization tasks in recommender systems. They followed a five-steps process for mining and weaving the taste fabric from social network profiles: (1) acquiring the profiles from social networking sites, (2) segmentation of the natural language profiles to produce a bag of descriptors of user characteristics, (3) mapping the natural language fragment descriptors into formal ontology, (4) learning the correlation matrix, and (5) discovering taste neighborhoods via morphological opening and labeling the network topology. By doing this it is possible to construct multi-faceted user profiles and find neighborhood of similar users over several domains, hence to enable cross-domain recommendations. They evaluated their approach by mining data from two online social networks and using it for recommendation in various combinations of their approach, including one that resembles classical collaborative filtering and showed that the use of their "taste fabric" provides better results. Seth and Zhang [74] used social networks data for building a user profile to be used for news stories recommendation. The main benefit from the ubiquitous user modeling point of view is that this information may be publicly available and accessible to the environment for bootstrapping an ad-hoc user model on the fly. The use of social networks for user modeling is a new dynamic and evolving area of research, especially due to easy access to freely available personal information in these networks (see also Refs. [76, 77, 80, 92, 95] and many more).

4.5 Mobile User Modeling

In ubiquitous computing, particularly in the case of the mobile user who needs a personalized service, we need to tackle the challenge of making sure that existing user models are available to the environment. Uhlmann and Lugmayr [82] describe the motivation for mobile profiles and provide an overview of the challenges and recent work on mobile personalization. To date, most of the personalized mobile applications developed have relied on server side user modeling, with the user model residing on the server while the user interacts with the system via a mobile device that sometimes captures also the user's context.

This mobile scenario calls for some additional tools beyond the use of a centralized user model server that may not be available/reachable every time the user needs it and everywhere. A possible solution is client-side personalization. In this case, the environment has to interact with the user (or the user's agent) to define the available services and the required personal data needed to drive them. For privacy protection, the user should be able first to decide what information she is willing to provide in return for the potential value of the personalized service. Client side personalization introduces new challenges for user modeling. First, there is a need to communicate with the environment, in order to negotiate and exchange user modeling data. Initial work in that area includes User ML, a user modeling language developed by Heckmann and Kruger [33]; this provided a first step towards bridging the gap between user modeling servers and occasional clients. Then there is a need to have a mobile user model, stored and maintain on a limited mobile device (both in storage and computing power). Heckmann and Kruger [33] provided an initial idea by pointing and referring to GUMO [34] – so the user model is a partial overlay on GUMO, which may help standardize the representation and interaction.

Additional work targeted specifically at personalization in mobile scenarios has been the focus of several researches. Myrhaug et al. [62] and Göker et al. [28] demonstrated a system that provided personalized and context aware information to users in ubiquitous computing environment by using special "context tags". The tags provided relevant information to users in their vicinity, while the context middleware on the device filtered the information based on the context of the user that included personal characteristics. In this case, the mobile context involved an integrated system, so avoiding some of the challenges of ubiquitous user modeling we have identified earlier. Quite a few similar systems appeared, providing personalized service to the user on the go. However, such stand alone systems do not provide any real solution for ubiquitous computing. Gonzáles et al. [29] went a step forward in their proposal for a generic multi-agent based user modeling framework. A user modeling agent may reside on a user mobile device or on a desktop and the user personal characteristics and the user modeling parameters can be mapped a priori to specific user modeling requirements for specific domains, represented by specific agents. Hence, whenever a service from an application is required, the application agent interacts with the user model agent to get the relevant information as needed.

Kuflik and Poteriaykina [52] proposed a user model on a personal device, where, the environment interacts with a personal agent on the device to provide a personalized service. The environment also negotiates with the agent to acquire the user modelling data needed/available for personalization. The mobile device must provide the data in the format required by the environment. The proposal combines a mobile personal server with a user model mediator and UserML based communication language for enabling exchange of user modelling in ubiquitous computing. Gerber et al. [26] created a somewhat different approach – PersonisJ a personal user model stored on the phone and based on Personis [47]. In their approach, the mobile phone uploads an application which provides the personalized service on the phone, with security and privacy controls ensuring that the user model remains on the phone; the application cannot send it off the phone. However, PersonisJ requires that the application be able to interpret and use the user modelling ontology and data.

4.6 A Word About Context Awareness

Context awareness plays a major role in ubiquitous computing. Notably, context includes the location and time, as well as other arbitrary information describing the user's current situation. Obviously, this can have an impact on personalization as well, meaning that the personalization should take account of it, regardless of the personalization approach or technique employed [79]. Even though context awareness is tightly coupled with user modeling, as discussed by Jameson and Krüger [41], context awareness is out of scope for this chapter. The reader is referred to a vast body of work on context awareness in ubiquitous or pervasive computing. One outstanding example is Dey et al. [19] who defined a generic framework for modeling and using contextual information. Other context aware surveys include [5, 36, 42, 78].

5 Discussion and Summary

Jameson and Krüger [41] implicitly noted the relationship between user modeling and ubiquitous computing. However, they correctly noted:

> We believe that the field of user modeling can contribute significantly to the enhancement of the effectiveness and usability of ubiquitous computing systems. In turn, the field of ubiquitous computing, by building the technological basis for mobile and migrating systems, is offering the user modeling community opportunities to apply their methods to novel types of systems, extending the methods themselves in the process.

However, so far very little research has taken this opportunity. Most solutions suggested so far have focused on narrow solutions. The main challenge presented by ubiquitous computing is the notion of "context" of the user that is complicated and dynamically changing and must be taken into consideration in addition to longer term, more stable user characteristics.

This chapter has provided a brief introduction to user modeling and then introduced ubiquitous user modeling, with its unique characteristics and challenges. We have outlined various partial solutions that exist or that are the subject of current research and offer promise for ubiquitous user modeling. It aimed to introduce the challenges as well as state-of-the-art partial solutions of ubiquitous user modeling. While there is a great variety of possible solutions, as briefly surveyed above, the main challenge is still unsolved – how to provide a personalized service for a first time user in ubiquitous computing? Addressing this challenge requires mutual effort of user modeling researchers and researchers working on ubiquitous computing – they must provide an integrated solution for personalization in ubiquitous computing, as so clearly described by Jameson and Krüger [41].

Ubiquitous display environments constitute a special case in ubiquitous computing because users approach and interact with situated public displays, seeking to benefit from a personalized service. To achieve this, we need to find ways to enable the environment to seamlessly acquire their relevant personal information (while respecting their privacy policies) and use it to provide a service matching their needs and preferences, without forcing them to explicitly interact to provide information to bootstrap the user model. Hence interaction between the environment and a mobile device seems the most promising approach for negotiating the service and the personal information. To achieve this, integration of the above-mentioned techniques is needed as well as addressing challenges related to user modeling representation – what information is available in a user model and how it is represented? How this information is made available to the environment, including the interaction between the user model and the environment are required. These are still yet to be resolved issues.

References

1. Abowd, G. D.: 1999, Software engineering issues for ubiquitous computing. In Proceedings of the 21st international Conference on Software Engineering (Los Angeles, California, United States, May 16–22, 1999). ICSE '99. ACM, New York, NY, 75–84.
2. Abowd, G. D., and Mynatt, E. D.: 2000, Charting past, present, and future research in ubiquitous computing. ACM Transactions on. Computer.-Human. Interaction. 7(1), 29–58.
3. Agnar, A., and Plaza, E.: 1994, Case-Based Reasoning: Foundational Issues, Methodological Variations, and System Approaches, Artificial Intelligence Communications 7(1), 39–59.
4. Assad, M., Carmichael, D., Kay, J., and Kummerfeld, B.: 2007, PersonisAD: Distributed, Active, Scrutable Model Framework for Context-Aware Services. Pervasive 2007, Toronto, Ontario, Canada, Springer, 4480, 55–72.
5. Baldauf, M., Dustdar, S., and Rosenberg, F.: 2007, A survey on context-aware systems. International Journal of Ad Hoc Ubiquitous Computing. 2(4), 263–277.
6. Berkovsky, S., Kuflik, T., and Ricci, F.: 2008, Mediation of user models for enhanced personalization in recommender systems. User Modeling And User-Adapted Interaction. 18 (3), 245–286.
7. Billsus, D., and Pazzani, M. J.: 2000, User modeling for adaptive news access. User Modeling and User-Adapted Interaction 10(2–3), Special Issue on Deployed User Modeling, 147–180.
8. Bohrer, K., and Holland, B.: 2000, Customer profile exchange (cpexchange) specification. http://www.cpexchange.org/
9. Brusilovsky, P.: 1996, Methods and techniques of adaptive hypermedia. User Modeling and User-Adapted Interaction 6(2–3), 87–129.
10. Brusilovsky, P., and Maybury, M. The Adaptive Web.: 2002, Communications of the ACM, 2002.
11. Brusilovsky, P., Sosnovsky, S., and Yudelson, M.: 2005, Ontology-based Framework for User Model Interoperability in Distributed Learning Environments. In G. Richards (ed.), Proceedings of World Conference on E-Learning in Corporate, Government, Healthcare, and Higher Education 2005, 2851–2855. Chesapeake, VA: AACE.
12. Burke, R.: 1999, Integrating Knowledge-based and Collaborative-filtering Recommender Systems. In Proceedings of the Workshop on AI and Electronic Commerce, at the AAAI 99 conference.
13. Burke, R.: 2000, Knowledge-based Recommender Systems, in A. Kent (ed.), Encyclopedia of Library and Information Systems. 69, Supplement 32.
14. Burke, R.: 2002. Hybrid recommender systems: Survey and experiments. User Modeling and User-Adapted Interaction, 12(4):331–370.
15. Cardoso, R. S., and Issarny, V.: 2007, Architecting Pervasive Computing Systems for Privacy: A Survey. In Proceedings of the Sixth Working IEEE/IFIP Conference on Software Architecture (January 06–09, 2007). WICSA. IEEE Computer Society, Washington, DC, 26.
16. Carmagnola, F., and Dimitrova, V.: 2008, An Evidence-Based Approach to Handle Semantic Heterogeneity in Interoperable Distributed User Models. In Proceedings of the 5th international Conference on Adaptive Hypermedia and Adaptive Web-Based Systems (Hannover, Germany, July 29–August 01, 2008). W. Nejdl, J. Kay, P. Pu, and E. Herder, Eds. Lecture Notes In Computer Science, vol. 5149. Springer-Verlag, Berlin, Heidelberg, 73–82.
17. Carmagnola, F., Cena, F., Console, L., Cortassa, O., Gena, C., Goy, A., Torre, I., Toso, A., and Vernero, F.: 2008, Tag-based user modeling for social multi-device adaptive guides. User Modeling and User-Adapted Interaction 18(5), 497–538.
18. Carmichael, D. J., Kay, J., and Kummerfeld, B.: 2005, Consistent Modeling of Users, Devices and Sensors in a Ubiquitous Computing Environment. User Modeling and User-Adapted Interaction 15, 3–4, 197–234.
19. Dey, A. K., Abowd, G. D., and Salber, D.: 2001. A conceptual framework and a toolkit for supporting the rapid prototyping of context-aware applications. Hum.-Comput. Interact. 16, 2 (Dec. 2001), 97–166.

20. Denaux, R., Dimitrova, V., and Aroyo, L.: 2004, Interactive Ontology-Based User Modeling for Personalized Learning Content Management, SW-EL'04: Semantic Web for E-Learning, workshop co-located with, AH 2004: International Conference on Adaptive Hypermedia and Adaptive Web-Based Systems, Eindhoven, The Netherlands, August 23–26.
21. Dolog, P., and Nejdl, W.: 2003. Challenges and benefits of the semantic web for user modelling. In Proc. of AH2003 – Workshop on Adaptive Hypermedia and Adaptive Web-Based Systems, WWW2003 Conference, Hypertext'03 Conference, User Modelling Conference 2003, Budapest, Hungary, Nottingham, UK, Johnstown, PA, USA, May 2003.
22. Dolog, P., Henze, N., Nejdl, W., and Sintek, M.: 2004, The Personal Reader: Personalizing and Enriching Learning Resources using Semantic Web Technologies. In Proc. of AH2004 – International Conference on Adaptive Hypermedia and Adaptive Web-Based Systems, August, 2004, Eindhoven, The Netherlands. Springer Verlag
23. Endres, C., and Butz, A. "A Survey of Software Infrastructures and Frameworks for Ubiquitous Computing", Mobile Information Systems Journal, 1(1), Jan–Mar 2005.
24. Fink, J., and Kobsa, A.: 2000, A Review and Analysis of Commercial User Modeling Servers for Personalization on the World Wide Web. International Journal on User Modeling and User-Adapted Interaction, 10(2–3), 209–249.
25. Fink, J., and Kobsa, A.: 2002. User Modeling for Personalized City Tours. Artificial Intelligence. Review 18(1), 33–74.
26. Gerber, S., Pink, G., Fry, M., Kay, J., Kummerfeld, B., and Wasinger, R.: 2010, PersonisJ: mobile, client-side user modelling, TR648, School of Information Technologies, University of Sydney.
27. Goldberg, D., Nichols, D., Oki, B. M., and Terry, D.: 1992, Using collaborative filtering to weave an information tapestry. Commun. ACM 35, 12, 61–70.
28. Göker, A., Watt, S., Myrhaug, H. I., Whitehead, N., Yakici, M., Bierig, R., Nuti, S. K., and Cumming, H.: 2004. An ambient, personalised, and context-sensitive information system for mobile users. In Proceedings of the 2nd European Union Symposium on Ambient intelligence (Eindhoven, Netherlands, November 08–10, 2004). EUSAI '04, vol. 84. ACM, New York, NY, 19–24.
29. González, G., Angulo, C., López, B., and de la Rosa, J. L.: 2005, Smart User Models for Ambient Recommender Systems, in Ambient Intelligence and (Everyday) Life, eds., Y. Cai. and J. Abascal, 113–122, San Sebastian, Spain, (July 2005). University of the Basque Country.
30. Gross, T.: 2008, Cooperative ambient intelligence: towards autonomous and adaptive cooperative ubiquitous environments. Int. Journal. On Autonomous. Adaptive. Communication. Systems. 1(2), 270–278.
31. Hanani, U., Shapira, B., and Shoval, P.: 2001, Information Filtering: Overview of Issues, Research and Systems. User Modeling and User Adapted Interactions, 11(3), 203–259.
32. Hatala, M., and Wakkary, R.: 2005, Ontology-Based User Modeling in an Augmented Audio Reality System for Museums. User Modeling and User-Adapted Interaction 15(3–4),339–380.
33. Heckmann, D., and Kruger, A. (2003) "A User Modeling Markup Language (UserML) for Ubiquitous Computing". In *9th International Conference on User Modeling (UM'2003)*, Pittsburgh, June, pp. 393–397
34. Heckmann, D.: 2005, Ubiquitous User Modeling, Ph.D. thesis, Computer Science Department, Saarland University, Germany.
35. Henricksen, K., Indulska, J., and Rakotonirainy, A.: 2001, Infrastructure for pervasive computing: Challenges. In GI Jahrestagung (1), 214–222.
36. Henricksen, K., Indulska, J., McFadden, T., and Balasubramaniam, S.: 2005, Middleware for distributed context-aware systems. In Proc. of On the Move to Meaningful Internet Systems, LNCS 3760, 846–863.
37. Henze, N., and Kriesell, M.: 2004, Personalization functionality for the semantic web: Architectural outline and first sample implementation. In Proceedings of the 1st International Workshop on Engineering the Adaptive Web (EAW 2004), co-located with AH 2004, Eindhoven, The Netherlands.

38. Henze, N., Krause, D.: 2006, Personalized access to web services in the semantic web. In: Cruz, I., Decker, S., Allemang, D., Preist, C., Schwabe, D., Mika, P., Uschold, M., Aroyo, L. (eds.) ISWC 2006. LNCS, vol. 4273, Springer, Heidelberg.
39. Iqbal, R., Sturm, J., Kulyk, O., Wang, J., and Terken, J.: 2005. User-centred design and evaluation of ubiquitous services. In Proceedings of the 23rd Annual international Conference on Design of Communication: Documenting & Designing For Pervasive information. Coventry, United Kingdom, September 21–23.
40. Izadi, S., Coutinho, P., Rodden, T., and Smith, G.: 2002, The fuse platform: Supporting ubiquitous collaboration within diverse mobile environments. Automated Software Engineering, 9(2), 167–186.
41. Jameson, A., and Krüger, A.: 2005. Preface to the Special Issue on User Modeling in Ubiquitous Computing. User Modeling and User-Adapted Interaction 15(3–4), 193–195.
42. Kjær, K. E.: 2007, A survey of context-aware middleware. In Proceedings of the 25th Conference on IASTED international Multi-Conference: Software Engineering (Innsbruck, Austria, February 13–15, 2007). W. Hasselbring, Ed. IASTED International Multi-Conference: Software Engineering. ACTA Press, Anaheim, CA, 148–155.
43. Kay, J.: 1990, um: a user modelling toolkit, Second International User Modeling Workshop, Hawaii, 251–261.
44. Kay, J.: 1994. Lies, damned lies and stereotypes: pragmatic approximations of users, Kobsa, A and D Litman (eds.), Proceedings of the Fourth International Conference on User Modeling, UM Inc, 175–184.
45. Kay, J.: 1995. The um toolkit for cooperative user modeling, User Modeling and User-Adapted Interaction 4(3), 149–196.
46. Kay, J.: 1999. Ontologies for reusable and scrutable student models, Mizoguchi, R, (ed) AIED Workshop W2: Workshop on Ontologies for Intelligent Educational Systems, 72–77.
47. Kay, J., Kummerfeld, B., and Lauder, P.: 2002, Personis: A server for user models. In: AH'02: Proceedings of Adaptive Hypermedia and Adaptive Web-Based Systems, Springer-Verlag, London, UK, 203–212.
48. Kobsa, A.: 1995, Editorial. Using Modeling and User-Adapted Interaction 4(2), Special Issue on User Modeling Shell Systems, iii^v.A.
49. Kobsa. A.: 2001, Generic User Modeling Systems. User Modeling and User-Adapted Interaction 11(1–2), 49–63.
50. Kobsa, A., and Fink, J.: 2006, An LDAP-based User Modeling Server and its Evaluation. User Modeling and User-Adapted Interaction 16(2), 129–169.
51. Kobsa, A., and Wahlster, W.: 1989, User models in dialog systems. Springer-Verlag.
52. Kuflik, T., and Poteriaykina, K.: 2009, User Model on a Key. In Proceedings of Hypertext 2009, Torino, Italy, June 29th–July 1st 2009, 371–372.
53. Kyriacou E. D.: 2009, Enriching Lifelong User Modelling with the Social e-Networking and e-Commerce "Pieces of the Puzzle". In: First and Seventeenth International Conference on User Modeling, Adaptation, and Personalization (UMAP2009)
54. Kyriacou, D., Davis, H., and Tiropanis, T.: 2009, Evaluating Three Scrutability and Three Privacy User Privileges for a Scrutable User Modelling Infrastructure, In: First and Seventeenth International Conference on User Modeling, Adaptation, and Personalization (UMAP2009) 22–26.
55. Lyytinen, K., and Yoo, Y. (eds.): 2002. Communications of the ACM 45 (12): 62–96.
56. Langheinrich, M.: 2001, Privacy by design – principles of privacy-aware ubiquitous systems. In Proceedings of UbiComp 2001, 273–291, Atlanta, GA, USA.
57. Lehikoinen, J. T., Lehikoinen, J., and Huuskonen, P.: 2008. Understanding privacy regulation in ubicomp interactions. *Personal Ubiquitous Computing* 12(8), 543–553.
58. Liu, H., Maes, P., and Davenport, G.: 2006, Unraveling the taste fabric of social networks. International Journal on Semantic Web and Information Systems, 2(1), 42–71.
59. Malone, T., Grant, K., Turbak, F., Brobst, S., and Cohen, M.: 1987, intelligent information sharing systems. Communications of the ACM 30(5), 390–402.

60. McCarthy, J.: 2001, Active environments: Sensing and responding to groups of people. Personal and Ubiquitous Computing, 5(1), Available at http://www.inf.ethz.ch/vs/events/dag2001/.
61. Mehta, B., Nieder'ee, C., Stewart, A., Degemmis, M., Lops, P., and Semeraro. G.: 2005, Ontologically-enriched unified user modeling for cross-system personalization. In User Modeling, 119–123.
62. Myrhaug, H., Whitehead, N., Goker, A., Faegri, T. E., and Lech, T. C.: 2004, AmbieSense – A System and Reference Architecture for Personalised Context-Sensitive Information Services for Mobile Users. 2nd European Symposium on Ambient Intelligence EUSAI 2004, Eindhoven, Netherlands, LNCS Volume 3295/2004 Springer-Verlag 327–338.
63. Mostefaoui, G. K., Pasquier-Rocha, J., and Brezillon, P.: 2004, Context-Aware Computing: A Guide for the Pervasive Computing Community. In Proceedings of the IEEE/ACS international Conference on Pervasive Services (July 19–23, 2004). ICPS. IEEE Computer Society, Washington, DC, 39–48.
64. Niemelä, E., and Latvakoski, J.: 2004. Survey of requirements and solutions for ubiquitous software. In Proceedings of the 3rd international Conference on Mobile and Ubiquitous Multimedia (College Park, Maryland, October 27–29, 2004). MUM '04, vol. 83. ACM, New York, NY, 71–78.
65. Niu, W. T., and Kay, J.: 2010, PERSONAF: Framework for personalized ontological reasoning in pervasive computing. User modeling and User-Adapted Interaction: the Journal of Personalization Research, 20(1) 1–40.
66. Pham, T., Schneider, G., and Goose, S.: 2000, A situated computing framework for mobile and ubiquitous multimedia access using small screen and composite devices. In Proceedings of the Eighth ACM international Conference on Multimedia (Marina del Rey, California, United States). MULTIMEDIA '00. ACM, New York, NY, 323–331.
67. Razmerita, L., Angehrn, A., and Maedche, A.: 2003, Ontology based user modeling for Knowledge Management Systems, Proceedings of the User Modeling Conference, Pittsburgh, USA, Springer Verlag, 213–217.
68. Ricci, F., and Werthner, H.: 2002, Case base querying for travel planning recommendation. Information Technology and Tourism, 3(3/4), 215–226.
69. Rich, E.: 1983, Users are Individuals: individualizing user. International Journal of Man-machine studies, 18(3), 199–214.
70. Saha, D., and Mukherjee, A.: 2003, Pervasive computing: a paradigm for the 21st century. Computer, 36(3), 25–31.
71. Sarwar, B., Karypis, G., Konstan, J., and Riedl, J. 2001. Item-based collaborative filtering recommendation algorithms. In Proceedings of the 10th international conference on World Wide Web (WWW '01). ACM, New York, NY, USA, 285–295.
72. Satyanarayanan, M.: 2002, Pervasive computing: vision and challenges. Personal Communications, IEEE, 8(4), 10–17.
73. Self, J.: 1990, Bypassing the intractable problem of student modelling, Intelligent Tutoring System: At the Crossroads of Artificial Intelligence and Education, 107–123.
74. Seth, A., and Zhang, J.: 2008, A Social Network Based Approach to Personalized Recommendation of Participatory Media Content. Proc. ICWSM'08. Seattle, WA.
75. Slay, H., and Thomas, B.: 2006, Interaction and visualisation across multiple displays in ubiquitous computing environments. In Proceedings of the 4th international Conference on Computer Graphics, Virtual Reality, Visualisation and interaction in Africa (Cape Town, South Africa, January 25–27, 2006). AFRIGRAPH '06. ACM, New York, NY, 75–84.
76. Song, X., Tseng, B., Lin, C., and Sun, M.: 2006. Personalized recommendation driven by information flow. SIGIR.
77. Stan, J., Egyed-Zsigmond, E., Joly, A. and Maret, P. "A User Profile Ontology For Situation-Aware Social Networking," 3rd Workshop on Artificial Intelligence Techniques for Ambient Intelligence (AITAmI2008), Jul. 2008.

78. Strang, T., and Linnhoff-Popien, C.: 2004, A context modeling survey. In 1st Int. Workshop on Advanced Context Modelling, Reasoning and Management.
79. Sutterer, M., Droegehorn, O., and David, K.: 'Upos: 2008, User profile ontology with situation-dependent preferences support', Proceedings of the First International Conference on Advances in Computer-Human Interaction, ACHI 2008. February 10–15, 2008 – Sainte Luce, Martinique, 230–235.
80. Torre, I.: 2009, Adaptive systems in the era of the semantic and social web, a survey. User Modeling and User-Adapted Interaction 19(5), 433–486.
81. Tutzschke, J., and Zukunft, O.: 2009, FRAP: a framework for pervasive games. In Proceedings of the 1st ACM SIGCHI Symposium on Engineering interactive Computing Systems (Pittsburgh, PA, USA, July 15–17, 2009). EICS '09. ACM, New York, NY, 133–142.
82. Uhlmann, S., and Lugmayr, A.: 2008, Personalization algorithms for portable personality. In Proceedings of the 12th international Conference on Entertainment and Media in the Ubiquitous Era (Tampere, Finland, October 07–09, 2008). MindTrek '08. ACM, New York, NY, 117–121.
83. van Elst, L., Abecker, A., and Maus. H.: 2001, Exploiting User and Process Context for Knowledge Management Systems. Workshop on User Modeling for Context-Aware Applications at the 8th Int. Conf. on User Modeling, July 13–16, 2001, Sonthofen, Germany.
84. Vassileva, J.: 1996, A task-centered approach for user modeling in a hypermedia office documentation system, User Modeling and User-Adapted Interaction, 6, 185–223.
85. Vassileva, J., Mccalla, G., and Greer, J.: 2003. Multi-Agent Multi-User Modeling in I-Help. User Modeling and User-Adapted Interaction 13(1–2), 179–210.
86. Walsh, E., Wade, V.:2009, Lowering the Barriers to User Model Interoperability, Ubiquitous User Modeling in proceedings of the Workshop at the Conference on User Modeling, Adaptation, Personalization (UMAP), Trento, Italy, 22–26 June 2009. 44–47.
87. Waller, V., and Johnston, R. B.: 2009. Making ubiquitous computing available. Communication of the. ACM 52(10), 127–130.
88. Weiser, M.: 1991, The computer for the 21st century, Scientific American, 3(265):94–014.
89. Weiser, M.: 1993, Some computer science issues in ubiquitous computing. Communications of the ACM 36(7), 75–84.
90. Weiβenberg, N., Voisard, A., and Gartmann, R.: 2004. Using ontologies in personalized mobile applications. In Proceedings of the 12th Annual ACM international Workshop on Geographic information Systems (Washington DC, USA, November 12–13, 2004). GIS '04. ACM, New York, NY, 2–11.
91. Wilensky, R., Arens, Y., and Chin, D.: 1984, Talking to UNIX in English: an overview of UC. Communications of the ACM 27(6), 574–593.
92. Yang, J., Wang, J., Clements, M., Pouwelse, J., de Vries, A. P., and Reinders, M.: 2007. An epidemic-based p2p recommender system. SIGIR-WS Large Scale Distributed Systems.
93. Yimam-Seid, D., and Kobsa, A.: 2003, Expert Finding Systems for Organizations: Problem and Domain Analysis and the DEMOIR Approach. Journal of Organizational Computing and Electronic Commerce 13(1), 1–24.
94. Yovanof, G. S., and Hazapis, G. N.: 2009. An Architectural Framework and Enabling Wireless Technologies for Digital Cities & Intelligent Urban Environments. Wireless. Personal. Communication. 49(3), 445–463.
95. Yu, B., and Singh, M.: 2003. Searching social networks. In Proceedings of the second international joint conference on Autonomous agents and multiagent systems (AAMAS '03). ACM, New York, NY, USA, 65–72.
96. Yudelson, M., Gavrilova, T., and Brusilovsky. P.: 2005, Towards user modeling metaontology. In L. et al. Ardissono, editor, 10th Int. Conf. on User Modeling, UM'05, 448–452, Edinburgh, Scotland, UK.
97. Zhang, F., Song, Z., and Zhang, H.: 2006. Web Service Based Architecture and Ontology Based User Model for Cross-System Personalization. In Proceedings of the 2006 IEEE/WIC/ACM international Conference on Web intelligence (December 18–22, 2006). Web Intelligence. IEEE Computer Society, Washington, DC, 849–852.

Context-Sensitive Display Environments

Florian Daiber, Antonio Krüger, Johannes Schöning, and Jörg Müller

Abstract In this chapter we will discuss the relatively new research area of context-sensitive Displays. The cheap availability of display spaces in combination with sensor technologies which allows to make inferences about the context of the display will enable a multitude of novel applications. However the foundations of these sensor-display systems are not well understood. Therefore we will discuss in this chapter the basic setup and a research methodology which has helped us to systematically approach the problems we faced during the design of several context-sensitive display environments. For this purpose we will present three research systems: the iDisplay-system, a public information system which uses extensively context-filtered information to inform students in University buildings, the MobiDic-system an context-sensitive advertisement management system for digital outdoor signage and our research on the Münster Multitouch Wall, which demonstrates novel ways of interacting with large displays in public environments.

1 Background and Motivation

In the last 5 years, we have witnessed a steep decline in display costs. The price of LCD panels has dropped more than 30% in the last 2 years. LCD panels larger than 50 in. have become affordable, and smaller display sizes are to become common. This development has lead to display panels being distributed widely in our daily environments. Larger LCD panels can be found in public spaces where the level of pedestrian traffic is high to

---

F. Daiber (✉) • A. Krüger • J. Schöning
Innovative Retail Lab, German Research Institute for Artificial Intelligence (DFKI), Campus D3_2, Stuhlsatzenhausweg 3, 66123 Saarbrücken, Germany
e-mail: florian.daiber@dfki.de; krueger@dfki.de; johannes.schoening@googlemail.com

J. Müller
Deutsche Telekom Laboratories, TU Berlin, Ernst-Reuter-Platz 7, 10587 Berlin, Germany
e-mail: joerg.mueller@tu-berlin.de

A. Krüger and T. Kuflik (eds.), *Ubiquitous Display Environments*, Cognitive Technologies, DOI 10.1007/978-3-642-27663-7_3,

© Springer-Verlag Berlin Heidelberg 2012

give information to the public, for example, in train stations or airports. Very prominent examples include large display installations in mega cities, such as New York City, Tokyo and Shanghai. Furthermore, dropping costs do also promote the usage of smaller displays. More and more retailers are starting to introduce electronic price tags on which the price information can be changed whenever necessary, which helps to reduce the cost of printing and of labour. Small displays can also be found in everyday objects, such as electronic toothbrushes and key rings.

In summary, the primary driver for replacing paper displays with their electronic counterpart is the reduced costs of replacement of information. Moreover, the cheaper the technology gets, the broader the scope of this replacement will be. The higher the information update costs for a traditional paper-based display are, the more likely it is that they will be replaced sooner rather than later. While displays of train and plane schedules today appear almost only in digital formats, we still see many traditional analogue advertisement boards. With the trend of falling display costs, it is very likely that the latter class will also be subject to replacement in the next couple of years. This will lead to certain environments where all information (regardless of its type) will be presented through a digital medium, and most of the media will be displayed digitally.

While these displays will be introduced purely to reduce the costs involved in the classical processes of information creation, distribution, and consumption, these emerging display environments will have capabilities which go far beyond the usages of their analogue ancestors. Being dynamic information media, which can display information at potentially high frame rates, they are able to react dynamically to the presence of users, and thus provide contextual information, such as display navigation hints and personalized advice. For example, a system of electronic price tags will be able not only to show the prices of and additional information about the product, but also highlight products that are especially relevant to a customer at the front of the shop's display shelves. In order to realize such services, the digital displays need to be combined with sensors, which are able to detect the context of the display, such as the presence of users and their identity, as well as an idea of environmental variables, such as temperature, noise level and time of day. This development will lead to what we call *context-sensitive display environments*. These display environments will be able to react to their physical environment and at the same time change the appearance of the environment itself.

This chapter reflects on this development and presents different display environments that we have used as research platforms to investigate their various aspects, including technical issues and also the question of how to improve interaction with display environments and their usability.

The next section will give a short overview of the research areas involved and the different disciplines that need to be involved in such a research initiative. Afterwards we will discuss three different types of display environment: the iDisplay system, the MobiDic system, and our work on large scale multi-touch surfaces. We will conclude with a short summary.

2 Key Properties and Research Areas

In 1991, Marc Weiser published his famous article about "The Computer for the 21st Century" [32]. His vision was that computers would become somehow *ubiquitous*. Weiser concretizes his vision as follows:

> Specialized elements of hardware and software, connected by wires, radio waves and infrared, will be so ubiquitous that no one will notice their presence. [34]

This means that some of them are getting to be an integral, invisible and unobtrusive part of the environment. Depending on size, Weiser defined ubicomp device classes (*tabs*, *pads* and *boards*). The latter class, namely *boards*, are defined as yard-sized displays with which the user can interact seamlessly. Seminal research on public or semi-public board-sized display was done at Xerox Parc in the late 1980s [33] starting with the LiveBoard [7].

Dey defines context-aware computing as follows:

> A system is context-aware if it uses context to provide relevant information and/or services to the user, where relevancy depends on the user's task.

According to this definition, *context-sensitive display environments* support users by tailoring the interaction and presentation to suit both the user and the display space. Based on this, the following key dimensions and requirements arise: Personalization, Spatial distribution, Sensors, Modalities, Resources, Privacy, and Deployments. *Personalization* aims at supporting individual users and must take into account their personal characteristics (including knowledge, preferences, etc.). These core aspects of personalization are mainly investigated in the field of user modeling and intelligent interface design. *Spatial distribution* takes interaction concepts for heterogeneous device configurations into account. In display environments various *Sensors* exist that have various requirements, including sensor fusion, situation modeling, and inferences. Additional *Modalities* arise in such environments that require multi- and cross-modal interaction approaches. *Resources* are a critical factor in ubiquitous computing and therefore there is a need for adaptation to technical and cognitive resources. *Privacy* is another critical issue in public or semi-public display environments and technologies are needed that respect privacy. *Deployments* of such display environments need to balance complexity and the benefit that might be achieved by analyzing longitudinal deployments.

These requirements show that many different research fields are involved in ubiquitous display environments research (see Fig. 1). Primary research areas are in the field of artificial intelligence including user-modeling, intelligent user interfaces, and ubiquitous computing. The supporting areas, i.e., human computer interaction (HCI) and interaction design, cognitive psychology as well as architecture and art, play notably important side roles.

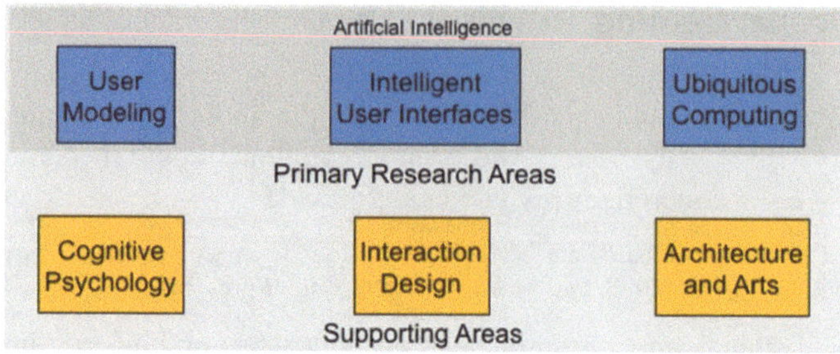

Fig. 1 Areas involved in context-sensitive display environments research

3 Public Displays

3.1 iDisplays: Self-Configuring Display Environments

Display environments will come in different flavors and technical settings depending on their size, distribution, and networking capabilities. The integration with sensors, such as cameras or web-based sensors will lead to displays that react to their environment and at the same time provide an adaptation of the environment. In this section we will discuss some interesting challenges associated with display environments. We will introduce the concept of news and reminder boards, which highlight different roles that information displays can play depending on their context of use. We will report from our experiences with a particular digital signage testbed, which has been installed at the University of Münster from 2005 until today. Another interesting problem that we will address is the question how to configure such display environments. Keeping in mind that certain display environments, such as a supermarket, will contain hundreds of displays, manual configuration will no longer be suitable.

Information displays and digital signage: Assuming that digital will, at a certain point in time, replace traditional signage, the question remains what type of information will be presented on this new category of displays in public spaces. Dynamic and timely information will especially benefit from the new flexibility of this medium. News and novelties can be presented and tailored to the personalized interests of the inhabitants of a particular display environment. Digital signage can be used for multiple purposes. A news display [19], which is usually used to provide novel information to an environment and is therefore public in nature, can be used temporarily for private tasks, such as finding a route. This blurs the traditional boundary of public and private displays. However, designers of such systems need to be careful to ensure that the information is well tailored to the needs of passers-by and inhabitants of the particular display environment. Research by Müller et al. [18] has shown that only if users assume that a display may contain information relevant to them are they willing to pay attention to the content. This is one of the major challenges ahead: to design the content of digital signage according to the expectations of users.

The iDisplays system: Digital displays offer a whole new way of presenting information in public spaces, essentially because the cost of changing information is so low. On most public displays that are installed nowadays, however, information is presented as slide shows or scrolling text. We argue that both these presentation formats are unsuitable for users who pass the displays often, because it is difficult for them to tell whether information has been updated. We propose using instead two different kinds of display, News Displays and Reminder Displays. Imagine the following scenario. A student passes the digital display installed at the entrance to his department (a News Display) every day. One day, as he has a quick look at it to see whether there is something new, he notices that there is an interesting talk next week, and notes it in his calendar. One week later, just before the start of the talk, a guest researcher who has just arrived passes the lecture hall. He sees the talk announcement on a different display (a Reminder Display) in front of the lecture hall and spontaneously decides to attend.

News Displays show new information chunks as soon as they are created. These displays support the process of planning well ahead, for example, for regular visitors to a place who see the displays often. Reminder Displays show chunks that are considered important at the current time and location by the system. They support the process of acting opportunistically, for example, for one-time visitors to a place. Note that the same information chunks are shown on News Displays when they are new and again on Reminder Displays when they are considered important. Thus many people will see chunks first on News Displays and then be reminded by Reminder Displays just in time.

In a first iteration, seven 40+ in. displays were rolled out in five lobbies of university buildings in Münster. Over 2,000 'users' pass by these displays regularly (predominantly students and university staff). Later on in the project phase more and more displays were installed step-by-step at additional university locations. The first prototype is based on a slide show paradigm showing a mixture of dynamic and hand-edited content (see Fig. 2).

IDisplays as an advanced research platform: Due to the promising results of the information and reminder displays in the first deployment phase of the project, more displays have been installed step-by-step at additional university locations. More than 22 displays are now operational in four different buildings. On the one hand, this leads to an increasing awareness and acceptance of such displays with the goal that more interaction with public displays will take place; on the other, it builds up a profound basis for research on public displays.

Learning topologies of situated public displays by observing implicit user interactions: How to measure context in a public display environment is one crucial aspect of setting up such environments. Assuming a display in such an environment has certain knowledge about its surrounding, it will be able to adapt its content according to the contextual knowledge. In Ref. [17], we proposed an approach to how such display environments can learn about their environment, especially their topological relationships, by implicit user interactions. This can be realized by the

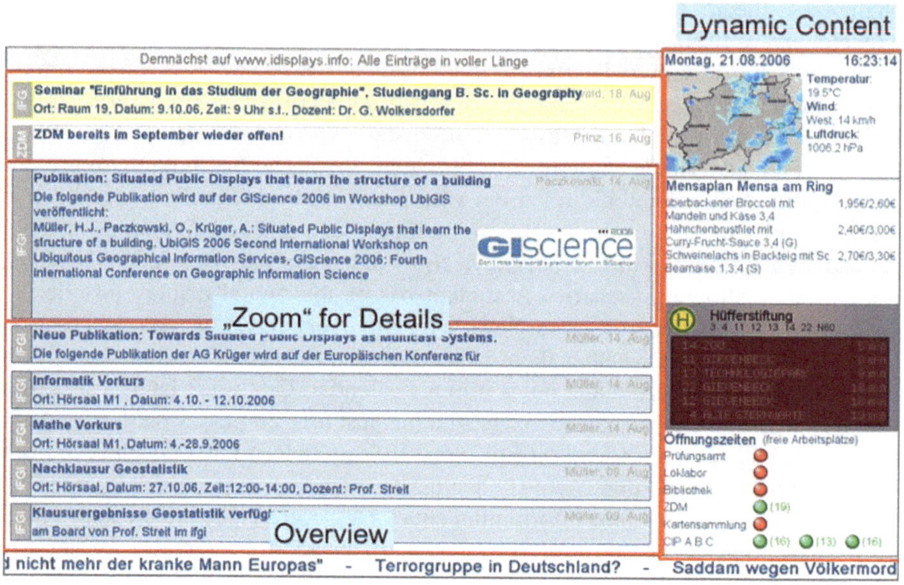

Fig. 2 iDisplay: information display showing a mixture of dynamic and hand-edited content

tracking of Bluetooth devices that users carry through the environment. Such a topological model stores only the relationship between the displays without the knowledge of their exact locations.

In our first approach, four displays covering two floors in a university building were used. The storage of "Enter" and "Leave" events of Bluetooth cells and transfer times between "Leave" and "Enter" events allows the identification of the shortest path between two displays.

To create a topological model of the displays, we follow a similar approach to that used in NearMe for long-range proximity [13]. In NearMe, to estimate the distance between two access points, the minimum travel time ever observed was used. This can be problematic in the case of frequent erroneous readings, for example, through ceilings and walls, and due to numerous other error sources. In this case, this approach is quite susceptible to the real distance being heavily underestimated. In our case, due to errors we sometimes have transit times of 0 or 1 s between displays, so taking the minimum would clearly produce erroneous results. Instead, we use the Expectation-Maximization (EM) algorithm [16]. This algorithm assumes that the different travel times are created by people taking different routes between the displays. It furthermore assumes that the travel times for each route are distributed according to the normal distribution. This is a strong simplification, because, for example, the normal distribution has non-zero probability for negative travel times while in reality negative travel times are impossible. Nevertheless, we were able to show that this approach still produces quite good results.

3.2 *MobiDiC: Context Adaptive Digital Ads on Public Displays*

Public space is increasingly penetrated by digital signs. Many stores use them for branding to improve the shopping experience, but increasingly digital signage is used to present advertisements from multiple advertisers. Such signs already exist in airports, train stations, and subway stations as well as in some public places, but will eventually replace paper signs everywhere. Because on digital signs it can be renewed every moment, the content can then be tailored to the context and the interests of the audience. Thus, scheduling becomes much more complex than for paper signs, and new technologies are needed to decide automatically which ad to show where and when. To enable such technology, the audience reaction to the content shown needs to be measured. Major challenges for digital signage will be the prevention of visual SPAM as well as the preservation of customer privacy. We hope that the approach presented here is a first step toward providing more interesting content on digital signs while preserving customer privacy.

In the MobiDiC project the interaction with public displays in conjunction with a couponing campaign was investigated. From October 2007 to the end of 2008 digital ads were displayed on German Telecom telestations. The users are enabled to interact with these displays via their mobile device. On the one hand, the user can take a photo of the displayed coupon to redeem the coupon in a (nearby) shop; on the other an advanced navigation service is provided that guides users to the intended shop. The pilot was deployed on 20 telestations, and 20 shops in downtown Münster participated. The architecture of the application is summarized in Fig. 3. Within this project it was investigated how users perceive digital ads on public displays through repertory grid analysis [12].

For digital signage, possibly many signs are controlled by a single system. On each of the signs, the content can be changed every moment. Thus, the scheduling problem of deciding where and when to show content for maximum effectiveness becomes too complex to be solved by humans.

Many mechanisms to allocate the display space automatically to multiple concurring advertisers are possible. Automatic auctions of software agents are a widely used approach to solving this kind of problem. One benefit is that pricing is natural. Each ad is represented by a software agent. The agent is provided with an advertising budget and can decide autonomously how much to bid in an advertising auction.

This rapid feedback loop is depicted in Fig. 4. At the beginning of an advertising cycle, the sign determines its context (Measure Context'), possibly time, location, weather, facial expression of audience or any other interesting context variables. This information is then provided to the advertising agents, which determine how much they should bid in the given context (Bidding Strategy'). The sign collects the bids and simply shows the content from the highest bidding agent (Show Content'). After showing the content, the sign determines the audience reaction to the content shown (Measure Audience Reaction), e.g. whether somebody actually redeemed the coupon (this information can be provided with a certain delay, however). From this information, the bidding strategies can then be refined, and the loop begins anew.

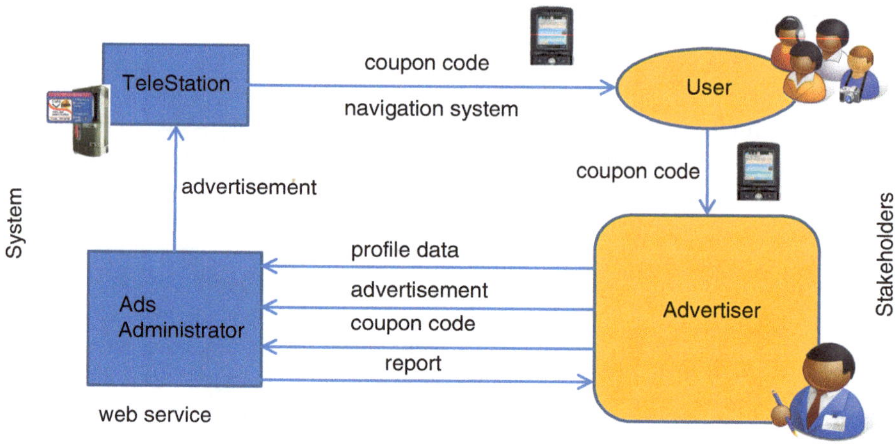

Fig. 3 Architecture of MobiDiC

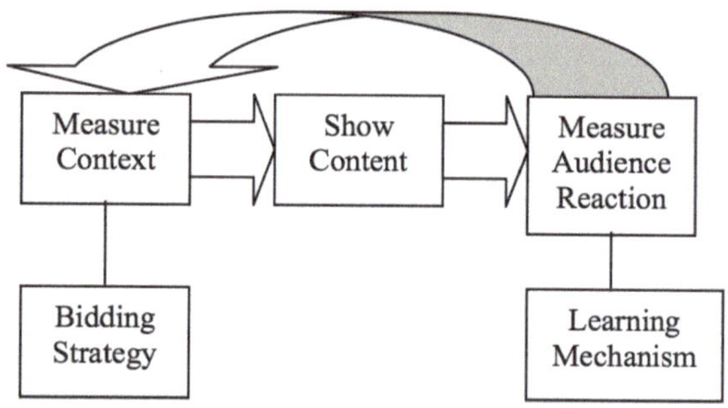

Fig. 4 The rapid feedback loop

3.3 *Multi-Touch Interaction with Wall-Sized Displays*

Multi-touch has great potential for exploring complex content in an easy and natural manner. Designers often make use of the geospatial domain to highlight the viability of their multi-touch interfaces because it provides a rich testbed. This is due to the fact that the command and control of geographic space (at different scales), as well as the selection, modification and annotation of geospatial data, are complicated [27]. One important observation in previous studies [11, 23, 25] is that users initially preferred simple gestures resembling mouse input of systems following the Windows-Icons-Menus-Pointer (WIMP) principle. After experiencing the potential of multi-touch, users tended towards physical gestures [35] to solve spatial tasks. Again users still prefer single hand gestures or gestures where the non-dominant hand just sets a frame of reference that determines the navigation mode, while the dominant hand specifies the amount of movement.

Multi-touch interactions with geospatial data: In order to provide a more natural interaction, we proposed a framework for manipulating spatial data on multi-touch

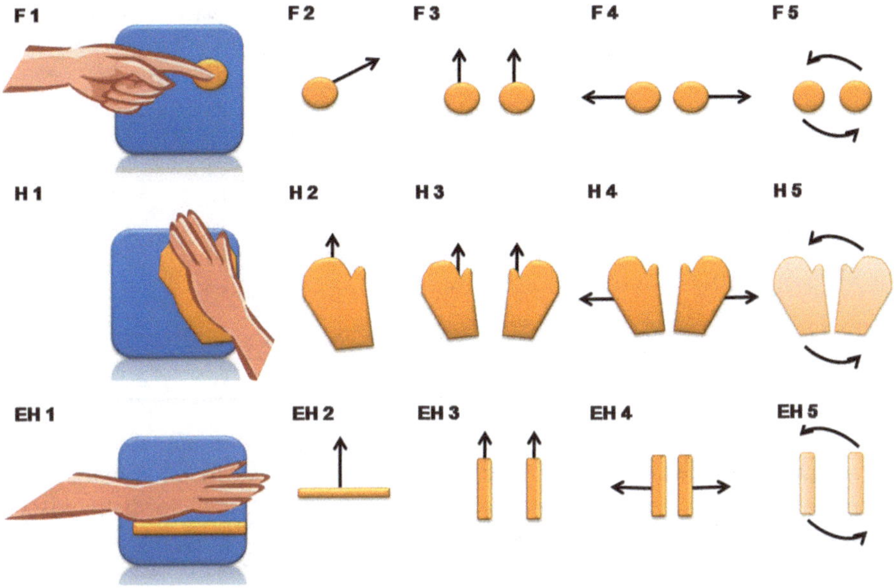

Fig. 5 Set of physical multi-touch gestures

enabled surfaces that rely on physical gestures in [5]. In contrast to the more general work of Wu et al. [38] and Wobbrock et al. [36], a conceptual framework for interaction with spatial data is presented. This framework is based on three key parts: physical interactions, interaction and manipulation of spatial data, and commands and controls that are needed to manipulate the geographic interaction space at different scales.

As a first step towards developing a framework for multi-touch interaction, various sets of simple physical interaction patterns were developed. Inspired by Wilson et al. [35] and Wu and Balakrishnan [37], a set of physical gestures for multi-touch input is derived (see Fig. 5). A set of interaction primitives is proposed that allows basic interaction tasks for geospatial data, such as pointing or zooming [27]. Commands and controls are designed that are needed to manipulate the geographic interaction space (at different scales) as well as to select, modify and annotate geo-objects. The interaction space is a simple set of graphical views and representations for spatial objects. Interaction primitives are used to modify the (graphical) view, e.g. zooming, panning, and rotating the globe (sphere) or plane (map view). It also contains feature manipulation, feature symbolization (simple symbol changes), feature filtering (hide a layer or feature, show another) and management of background images.

This results in a framework for physical multi-touch interaction with geospatial data that are summarized in Table 1. The rows represent the most common commands that are needed for geospatial tasks: point, zoom, pan, rotate, tilt (i.e. to control the parameters of a geographical view), and cut (as a representative for a

Table 1 Framework for physical multi-touch interaction with geospatial data (F2) instead of just pointing to it. This helps to reduce the ambiguity of the gesture as pointed out in Ref. [30]

	World		(Geo-) Objects			Symbols		
	Globe	Plain	Point	Line	Polygon	Point-symbols	Labels	Layer
POINT	F1	F1	F1	EH1	H1	F1	F1	F1
ZOOM	F4	F4	–	F4	F4	F4	F4	(F4)
PAN	H2	H2	F2	EH2	H2	F2	F2	–
ROTATE	F5	F5	–	F5	F5	F5	F5	–
TILT	H1 + F2, H1 + H2	H1 + H2	–	–	–	F6	–	–
CUT				EH1,EH3	EH1, EH3			–

topological operation such as dividing an area). The columns of the table represent the geographic space and objects that can be subjects of the various commands. The interaction (selection and manipulation) with geo-objects can be distinguished according to their geometric properties: point, line, and polygon. Finally, in geospatial applications one often finds interaction patterns with symbolic representations (such as a point-of-interest, in the table denoted as point-symbols), or their annotations, which we refer to as labels. Similar symbols are often organized in layers, which can themselves be the subject of manipulation. Interestingly, the geometric property of the interaction is reflected in the physical nature of the proposed multi-touch interaction. For example, single point-like objects are referred to with a single pointing gesture (F1), while rotation of a globe or panning of a 2D map is more likely to be performed by a wiping-style gesture (H2). The selection of geo-objects can be improved by referencing their geometric properties. For example, the selection of a street on a map could be more precisely performed by moving a finger along that street

Multi-user multi-touch interaction with public interactive displays: Large interactive screens installed in public spaces are not so far from fiction. Jeff Hans YouTube demonstration at the TED conference[1] captured the imagination of researchers and users alike. Technologies such as Frustrated Total Internal Reflection (FTIR) and Diffused Illumination (DI) [24] allow the low-cost development of such surfaces and have led to a number of technological and application innovations.

From an HCI perspective it is interesting to investigate how these large interactive screens can support different kinds of collaborative interactions. In order to study this we installed a large-scale multi-touch screen in a pedestrian underpass in the city of Münster for 1 week during a technology exhibit called "Germany Land of Ideas" (here also referred to as "Hightech Underground") (Fig. 6). Users were able to navigate through the world on a virtual globe and receive additional

[1] http://www.youtube.com/watch?v=QKh1RvOPlOQ.

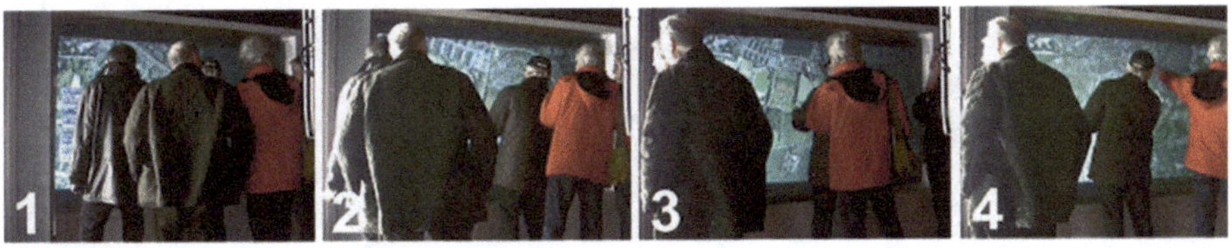

Fig. 6 A group of users interacting with a large-scale multi-touch display in a pedestrian underpass in the city of Münster, Germany

information about 365 Points of Interest (POI) in the "Land of Ideas". To our knowledge, this is the second attempt to analyze the interaction at a large multi-touch display in (or in our case under) a city center. In the spirit of the criticism of Greenberg and Buxton [9] and their claim of "Lack of Replication" we believe that our work provides data to verify and deepen findings of the sort that Peltonen et al. [20], Vogel and Balakrishnan [29], Prante et al. [21] and our own [25] surveys have presented previously.

Peltonen et al. [20] presented detailed observations from their own large multi-touch display, called CityWall, in the city of Helsinki, Finland. On the DI multi-touch display a photo explorer was presented. They observed various interaction patterns relating to how people used the CityWall installation and how they collaborated and interacted with each other in front of the display. Prante et al. [21] proposed different interaction zones (Cell Interaction Zone, Notification Zone, and Ambient Zone) for the interaction with a large public display. This work was followed up by Vogel and Balakrishnan [29] with their work on an interaction framework for sharable, interactive public ambient displays. Vogel et al. defined four interaction phases, facilitating transitions from implicit to explicit interaction and from public to personal interaction (personal interaction zone, subtle interaction zone, implicit interaction zone and ambient zone). In our own work [25] we focus on investigating the usage and user needs of virtual globes. We found that the main motivation of around half of the users (53.4% $\pm$ 11.6%) was to use a virtual globe for either looking at their own house or other individual places (e.g. a neighbor's house, their hotel from their last vacation, the city center).

An observational study was conducted in the "Hightech Underground" for 1 week. It was observed that the visitors got insights about spontaneous and group interaction patterns. The video[2] presents a short impression of the installation. The ambient nature of this wall setup does not attract users by itself. Users who watch other people interacting with the application before they interact seem to be less shy and more experimental [3]. Users had fun performing various gestures: rotating, flipping, and scaling the digital globe. This "fun-factor" played an important role due to the fact that most users were not that interested in the information presented for this exhibition. They were rather interacting with the globe in order to look, for example, for their houses, vacation houses or tennis court (see Fig. 3). Although designed for

[2] Hightech-Underground Installation: http://www.youtube.com/watch?v=27oSoDdU2fk.

Fig. 7 Dyads of users interacting with a large-scale multi-touch display in a pedestrian underpass in the city of Münster, Germany

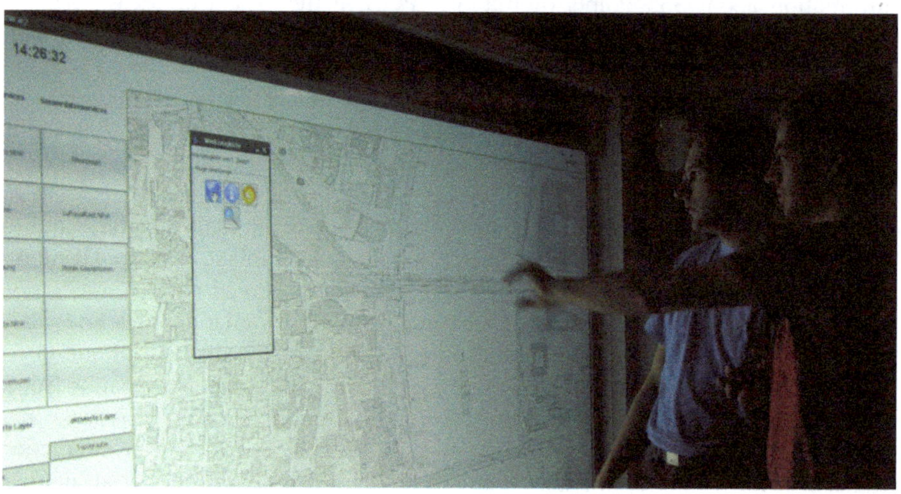

Fig. 8 Collaborative interaction in a crisis response scenario

single users, the application was mainly used by groups of two to eight people. An exemplary group of elderly men reveal this highly interactive group communication in Figs. 1, 2, 3, and 4. Thus, user groups of different sizes are able to interact collaboratively with the single-user application.

The formerly described "Teacher-apprentice setting" [20] also plays an important role in this study. Examples of this are shown in Fig. 7. In many cases one person (teacher) shows or explains something to "the student". Another collaborative interaction is shown in Fig. 8. A team of users performs parallel zoom-gestures.

Our preliminary investigation of the data we have collected largely verifies findings from literature. We could find similar interaction patterns as Peltonen et al. [20] namely: "huge amount of teamwork" (consider Figs. 1, 2, and 3), "Leaving a mark" and "Teacher-apprentice setting" (consider Fig. 4).

Multi-touch GIS interaction to support collaborative work in crisis response centers: Many approaches to improving crisis response systems exist but they are often focusing on standardization attempts for the integration of different information sources [28, 31] as well as on a trend toward Service-Oriented Architectures (SOA), e.g. the redevelopment of the German crisis-response information system (deNIS) to the service oriented version deNIS IIplus [2] (Fig. 8). Beside these strategies, the field of human computer interaction is another important research area in disaster management [4]. Early work in this field was done by Florence et al. [8] and Maceachren et al. [15].

Large multi-touch screens installed in crisis response centers improve the work of crisis response units in many ways, mainly by allowing intuitive multi-touch interaction with seldom used complex expert systems, and are open for real multi-user interaction allowing many users to work collaboratively and in parallel with the system. Therefore, the critical issue in crisis response scenario is the need for intuitive, easy-to-learn and easy-to-use software. Multi-touch interaction is well suited to fulfilling these requirements because it enables a more direct interaction with a direct link to the physical world (e.g. pointing or grabbing virtual objects).

As part of the SoKNOS project,[3] it was investigated how collaborative work at large multi-touch screens can be used to support target-oriented and adequate decisions in crisis response centers. The multi-touch wall is mainly used to get a quick overview of the up-to-date state of the crisis that can be discussed collaboratively in front of the wall. Therefore, a prototype was developed that provides a set of intuitive, easy-to-learn multi-touch gestures and enables multi-user interaction. Observations are made similar to the "Hightech Underground" study (see above). There is one interaction scheme where presentations are given in the "Teacher-Apprentice-Setting" style but there is also a parallel interaction scheme where different users are interacting with the wall at the same time, e.g. one is browsing through layers of information while another one is navigating on the map to an operation site.

4 Mobile Personalized Interaction with the Environment

Today a lot of geo-referenced information exists in electronic form. Based on this information, electronic maps and location-based services can be built that support navigation in unfamiliar areas or provide up-to-date information (e.g. events, services, etc.). Mobile devices equipped with additional sensors (GPS, compass etc.) can communicate with such location-based services and represent this geo-referenced information on a map. Unfortunately, the presentation of complex information on mobile devices is limited by the small display size. Thus using mobile devices for map representations has certain drawbacks. Due to the size and resolution of the device only low level detail can be represented on the display. It is therefore difficult to get an overview and hard to scroll through the map.

While public displays in general lack personalization, small personal devices allow personalized interaction with the environment. In the following section various approaches are presented that enable the user to interact with the environment through mobile phones and mobile projectors.

[3] http://www.soknos.de.

4.1 Wikeye

In the following section a combination of the advantages of mobile devices with the advantages of paper maps is presented. Using the mobile device like a "magic lens" (Video-See-Through device) over the map allows a variety of novel digital map interactions in a broad spectrum of application areas (e.g. tourist information, flight/ train schedules, consumer information, shopping, and advertisement).

Maps are turning from static information products printed on paper to dynamic representations of space displayed on various kinds of interactive electronic device (such as mobile devices or large scale interactive displays). Today, paper maps are still superior in some categories to their digital counterparts. They provide high resolution with zero power consumption. Because of the advantages of paper maps and the drawbacks of visualizing maps on mobile devices (small display, low resolution), the combination of mobile devices with external paper maps is an active research area.

With the Wikeye [10] application the user can sweep the mobile device display over paper maps and can augment the map with additional (personal or dynamic) information. We showed in Ref. [22] that this physical movement of the user's arm over the map significantly helped users to perform the task. Using small physical movements of the user's arm with the mobile device can also significantly improve the interaction.

Wikeye aimed at enhancing the understanding of places through the combination of digital Wikipedia content with a paper-based map. The goal of Wikeye is to help users to understand more about their surroundings via an easy-to-use mobile interface. When the user views a small portion of a map through her mobile device, Wikipedia-derived content relating to spatial objects in view is offered to the user. For example, when a Wikeye user examines a small portion of a Berlin paper map through the camera phone, such as the area containing the Reichstag building and the Brandenburg Gate, Wikipedia content is overlaid on the camera image of the map, highlighting these spatial objects and their relationships. Rotating the device about the camera axis enables an intuitive time browsing functionality following a clock metaphor. This allows the user to switch through time in history and retrieve an overview of content related to that time period. Depending on the spatial extent of the map visible on the camera display unit, the system responds by offering Wikipedia data on spatial objects with larger area footprints. In other words, as the cartographic scale decreases, the threshold area of a spatial object to be featured with Wikipedia data increases.

4.2 Photomap

In many mid- to large-sized cities public maps are ubiquitous. One can also find a great number of maps in parks or near hiking trails. Public maps help to facilitate orientation and provide special information not only to tourists but also to locals who just want to look up an unfamiliar place while on the go. These maps offer many advantages compared to mobile maps from services such as Google Maps

Mobile or Nokia Maps. They often show local landmarks and sights that are not shown on standard digital maps. Often these 'You are here' (YAH) maps are adapted to a special use case, e.g. a zoo map or a hiking map of a certain area. Being designed for a fashioned purpose, these maps are often aesthetically well designed and their usage is therefore more pleasant.

One approach to filling this gap is the novel and innovative[4] technique and application called PhotoMap [26] that uses snapshots of public maps in combination with GPS for ad hoc navigation. We give a brief overview of the scenario, discuss the properties of YAH maps, and present two approaches that support the user to georeference the taken image properly with sufficient accuracy to support pedestrian navigation tasks.

The following scenario illustrates the concepts more generally. Maximilian is walking through the city of Münster, Germany. He approaches the botanic garden behind the castle of the city. Standard maps of this area just show a grey (not even a green) area, but fortunately there is a paper YAH map at the entrance of the garden containing all details and POIs of the garden. Maximilian takes a picture of the map of the botanic garden with his GPS-enabled phone and performs some additional actions to do the georeferencing in an easy and appealing way. For this, we utilize the current GPS positions where Maximilian took the photo in order to establish the map's scale and extent (and northing).

The georeferencing step needs to be supported by the users themselves. Therefore, we developed two approaches explaining a trade-off between user interaction and system complexity, namely the "Two Point Referencing" and the "Smart Alignment" approach. In the first version, the map is rectified in two steps and the user has to move physically between these steps before being able to use the map. Our second implementation allows the user to adjust his photo of a YAH map to an already referenced map on her GPS trace and then use it as a foundation for navigation.

Generally, the projection of the map and the image distortion (caused by the tilting angle between map and mobile device) should not be too extreme. An optimal photo would be taken by a mobile device held parallel to the map (see Fig. 9). Reference points (assigning real world coordinates to the image pixel) need to be given by the user and need to be combined with the current GPS-coordinates provided from the GPS-module of the mobile phone. Currently we have implemented these two approaches for our PhotoMap prototype.

Two Point Referencing is a method to georeference the image by simply adding to reference points. Directly after the user has taken the image of the map she should either indicate where she is on the map or mark the YAH dot to determine her actual position. This marked position will then be synced via GPS. Then she has to move a significant distance and add another mark in a second step by simply repeating the process. After that PhotoMap rectifies the map and allows her to look up her location or even create a GPS trace or record a GPS track (see Fig. 9, *right*).

[4] Finalist of Nokia contest, "Calling all Inventors", with more than 1,000 competitors.

Fig. 9 Photo map interaction (*left*) and Two Point Referencing: shoot a photo of the map; set first reference point; walk 25–200 m; set second reference point (*right*)

She is able to pan around the map and can obtain either her actual position and/or her GPS trace. In the actual implementation the screen of the mobile device is about four times larger than the photo of the map. This is a trade-off between photo resolution and map scale.

Smart alignment, in contrast, is a georeferencing approach that involves fewer user interventions. After taking a photo of a YAH map, the user has to indicate again where she is on the map or mark the YAH dot to determine the actual user's position. Afterwards the photo of the map is displayed semi-transparently over the map data. The YAH dot is anchored to one corresponding position on the map. From our map corpus we utilise the knowledge that the scale is normally about 1:1,500 to 1:2,500 and that about 81% of the maps have the correct northing (see below). Hence the photo is roughly adjusted depending on the scale. The user has to perform the fine adjustment of the photo on the map (primarily scale and rotation). A slightly modified version of the "smart alignment" method utilizes the GPS trace of the user. Instead of overlaying the photo of the map on a digital map, the GPS trace is displayed. Of course, here it is assumed that the user's trace overlaps with the spatial extent of the photo of the map. As we can conclude from our map corpus, this is the case in nearly half of the cases (about 48%). Problems will occur if the YAH dot is positioned at an edge of the map.

As already stated above, the question remains whether YAH maps in general have the common properties that allow this generic approach for maybe all existing YAH maps. In order to explore this, we collected a corpus of 93 maps in 21 cities. These maps are of a variety of types (city maps, local areas, mountain areas, parks, etc.) and for many different purposes (sightseeing, shopping, hiking, etc.) but the interesting results are that about 96% of the maps were spatially correct, meaning they were designed using underlying map data and were of scales ranging from 1:1,500 to 1:2,500. 81% had the correct northing and 62% of the maps had a YAH dot. The distribution of the YAH dots is restricted to a few distinct areas on the map: in the middle of the map, the lower middle, and the upper left.

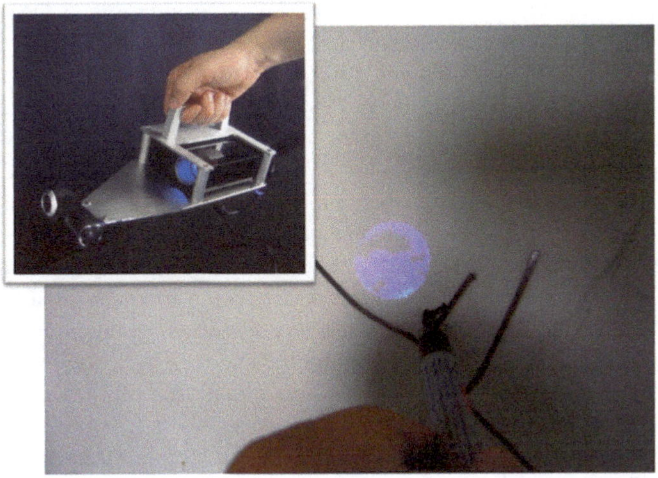

Fig. 10 LittleProjectedPlanet prototype and projection on a whiteboard sketch

4.3 LittleProjectedPlanet

In order to explore the interaction of mobile projection units with the environment, the LittleProjectedPlanet [14] prototype was developed. LittleProjectedPlanet is a mobile adaption of the Playstation 3 (PS3) game LittleBigPlanet. The camera projector unit is used to augment hand drawings of a user with an overlay displaying the physical interaction of virtual objects in the real world. Therefore a spatial synchronous projection setup is needed. Players can sketch a 2D world on a sheet of paper or use an existing physical configuration of objects and let the physics engine simulate physical procedures in this world to achieve game goals (see Fig. 10). With LittleProjectedPlanet we propose a mobile game combining hand drawn sketches of a user combined with objects following a physics engine to achieve certain game goals. Enriching sketching in combination with physical simulation (ASSIST) was presented by Davis et al. [1, 6].

LittleProjectedPlanet is a mobile augmented reality game for mobile camera projector units that integrates concepts of ASSIST (sketching game worlds on real world objects) and LittleBigPlanet ("play with everything" in a natural way using physics). More precisely the game enables the user to interact with a sketched 2D environment the user designed in the real world. So LittleProjectedPlanet is instantly playable in nearly every environment and no additional hardware is needed to set up the game. The camera detects and calculates a physical model from the environment (sketches or edges of physical objects). Into this model the user can place several virtual objects representing items such as tennis balls or bowling balls. These virtual objects then get projected into the real world by the mobile projector. When starting the physic engine, the application simulates the interaction of the virtual and the real world objects and projects the results of the virtual objects onto the real world surface.

Just as in LittleBigPlanet our application offers the user different ways of playing. One is like the level designer in LittleBigPlanet; the user can freely manipulate the 2D world within the projected area and place virtual objects in it. Similarly to children building tracks for marbles in a sandpit, the player can specify a route and then let the virtual marbles run along it. A different gaming mode is a level-based modus, but instead of steering a character as in LittleBigPlanet, the user designs the world by herself. As a goal the user has to steer a virtual object, e.g. a tennis ball, from its starting point to a given finish. The game concept uses a direct manipulation approach. Enabling the player to modify the world at runtime lets the real world objects become the user's tangible interface. But not only are the objects used for the interface by changing the orientation and position of the projector; the user can also modify the physical procedures (e.g. gravity by turning the mobile camera projector unit).

Figure 10 shows how a user is projecting virtual marbles on a track she sketched on a whiteboard. To allow the user a smooth and seamless interaction, the gravity in the projection is aligned with the real world's gravity. Also gravity can be utilized in the game to control some action. A user can take control of the gravity by changing the orientation of the projector. Doing this, the user can let virtual objects fly through the levels.

Due to the unavailability of sophisticated projector phones (with an optimal alignment of camera, built-in projector, and a GPU that is able to process the physics simulation) we used a Dell M109S, a mobile projector with a maximum resolution of 800×600 pixels and a weight of 360 g, in combination with a Logitech QuickCam 9000 Pro. Altogether our prototype weighs around 500 g and is therefore feasible to handle (see Fig. 10). For controlling the application and to determine the orientation (to set the gravity) a Nintendo Wii remote is attached to the camera projector unit.

The application is fully implemented in Java using the QuickTime API to obtain a camera image. A physics engine, Phys2D, an open source, Java-based engine, is used. The communication with the Wii remote is handled by Wii Remote.[5] Connected to a standard laptop or PC, the camera projector unit has a refresh rate of approximately 15 fps when running the application. The area of the camera image containing the projected image is processed via an edge recognition algorithm. Every pixel of a detected edge gets a representation as a fixed block in the physics engine. This gives the user total freedom in designing the world. Such a physic world update is done every 300 ms but it can be stopped by the users, for example for editing the sketch. The gravity of the physical model is adapted to the actual orientation of the camera projector unit through calculating the roll (this denotes the angular deviation along the longest axis) of the Wii remote.

---

[5] http://www.world-of-cha0s.hostrocket.com/WiiRemoteJ/.

5 Conclusions and Outlook

In this chapter we have presented different prototypical implementations of context-aware display systems. We have given an overview on the longitudinal deployment of the IDisplay system, which has provided context-aware information to inhabitants of university buildings. We have discussed various design decisions and presented our experiences with the overall system. Afterwards we have presented a large scale deployment of the digital advertisement system MobiDiC, which provides a platform to circulate digital coupons and addresses the specific needs of small enterprises. By auctioning the coupons on the basis of context information (such as weather or time), the coupons can be tailored much better to the various situations and are therefore also applicable for smaller marketing budgets. Finally, we have discussed our work on large-scale multi-touch surfaces that have been designed to provide spatial information in a flexible manner. We are currently continuing this work and we are looking deeper into highly connected environments where personal devices connect to a specific instrumented environment (e.g. a user's smart phone to the data warehouse of a shop) to provide sophisticated device spanning services on multiple displays, extending the current display islands to a true large-scale display environment. One particular scenario in which we are interested is that of personal navigation, where a portable projector is used in combination with a public display system to provide route instructions. Another challenge is issues that arise when multiple users are simultaneously interacting in display environments. Novel tracking and interaction techniques will be necessary to provide in such environments a seamless and satisfying user experience.

References

1. Alvarado, C., Davis, R.: SketchREAD: A Multi-Domain Sketch Recognition Engine. In: Proceedings of UIST 2004. pp. 23–32. ACM (2004)
2. BBK (Bundesamt für Bevölkerungsschutz und Katastrophenhilfe): deNIS IIplus – Rechnergestütztes Krisenmanagement bei Bund und Ländern. (2006)
3. Brignull, H., Rogers, Y.: Enticing people to interact with large public displays in public spaces. In: Proceedings of INTERACT '03. pp. 17–24 (2003)
4. Carver, L., Turoff, M.: Human-computer interaction: the human and computer as a team in emergency management information systems. Commun. ACM 50, 33–38 (March 2007)
5. Daiber, F., Schöning, J., Krüger, A.: Whole Body Interaction with Geospatial Data. In: Butz, A., Fisher, B., Christie, M. (eds.) Smart Graphics. vol. 5531, pp. 81–92. Springer (2009)
6. Davis, R.: Sketch Understanding in Design: Overview of Work at the MIT AI Lab. In: Sketch Understanding, Papers from the 2002 AAAI Spring Symposium. pp. 24–31. AAAI Press (2002)
7. Elrod, S., Bruce, R., Gold, R., Goldberg, D., Halasz, F., Janssen, W., Lee, D., Mc-Call, K., Pedersen, E., Pier, K., Tang, J., Welch, B.: Liveboard: a large interactive display supporting group meetings, presentations, and remote collaboration. In: CHI '92: Proceedings of the SIGCHI conference on Human factors in computing systems. pp. 599–607. ACM (1992)

8. Florence, J., Hornsby, K., Egenhofer, M.: The GIS wallboard: interactions with spatial information on large-scale displays. International Symposium on Spatial Data Handling 7, 449–463 (1996)
9. Greenberg, S., Buxton, B.: Usability evaluation considered harmful (some of the time). In: CHI '08: Proceeding of the twenty-sixth annual SIGCHI conference on Human factors in computing systems. pp. 111–120. ACM (2008)
10. Hecht, B.: Wikeye – using magic lenses to explore georeferenced wikipedia content. In: PERMID '07: Proceedings of the 3rd International Workshop on Pervasive Mobile Interaction Devices (2007)
11. Hornecker, E.: "I don't understand it either, but it is cool" – Visitor Interactions with a Multi-Touch Table in a Museum. In: TABLETOP '08: 3rd IEEE International Workshop on Horizontal Interactive Human Computer Systems. pp. 113–120 (2008)
12. Jankowicz, D.: The Easy Guide to Repertory Grids. Wiley (2003)
13. Krumm, J., Hinckley, K.: The nearme wireless proximity server. In: Davies, N., Mynatt, E., Siio, I. (eds.) UbiComp '04: Ubiquitous Computing, Lecture Notes in Computer Science, vol. 3205, pp. 283–300. Springer (2004)
14. Löchtefeld, M., Schöning, J., Rohs, M., Krüger, A.: LittleProjectedPlanet: An Augmented Reality Game for Camera Projector Phones. In: MIRW '09: Workshop on Mobile Interaction with the Real World (at MobileHCI 2009) (2009)
15. Maceachren, A.M., Brewer, I., Cai, G., Chen, J.: Visually-enabled geocollaboration to support dataexploration decision-making. In: Proceedings of the 21st International Cartography Conference. pp. 394–401 (2003)
16. Mitchell, T.: Machine Learning. McGraw-Hill Education (ISE Editions) (1997)
17. Müller, H.J., Krüger, A.: Learning topologies of situated public displays by observing implicit user interactions. In: UAHCI'07: Proceedings of the 4th international conference on Universal access in human-computer interaction: ambient interaction. pp. 158–167. Springer (2007)
18. Müller, J., Jentsch, M., Kray, C., Krüger, A.: Exploring factors that influence the combined use of mobile devices and public displays for pedestrian navigation. In: NordiCHI '08: Proceedings of the 5th Nordic conference on Human-computer interaction: building bridges. pp. 308–317. ACM (2008)
19. Müller, J., Krüger, A.: Mobidic: Context adaptive digital signage with coupons. In: AmI '09: Proceedings of the European Conference on Ambient Intelligence. pp. 24–33. Springer (2009)
20. Peltonen, P., Kurvinen, E., Salovaara, A., Jacucci, G., Ilmonen, T., Evans, J., Oulasvirta, A., Saarikko, P.: It's mine, don't touch!: interactions at a large multi-touch display in a city centre. In: CHI '08: Proceeding of the twenty-sixth annual SIGCHI conference on Human factors in computing systems. pp. 1285–1294. ACM (2008)
21. Prante, T., Röcker, C., Streitz, N., Stenzel, R., Magerkurth, C.: Hello.wall – beyond ambient displays. In: Ubicomp '03: In Adjunct Proceedings of Ubicomp. pp. 277–278 (2003)
22. Rohs, M., Schöning, J., Raubal, M., Essl, G., Krüger, A.: Map navigation with mobile devices: virtual versus physical movement with and without visual context. In: ICMI '07: Proceedings of the 9th international conference on Multimodal interfaces. pp. 146–153. ACM (2007)
23. Ryall, K., Forlines, C., Shen, C., Morris, M.R., Everitt, K.: Experiences with and observations of direct-touch tabletops. In: TABLETOP '06: Proceedings of the First IEEE International Workshop on Horizontal Interactive Human-Computer Systems. pp. 89–96. IEEE (2006)
24. Schöning, J., Brandl, P., Daiber, F., Echtler, F., Hilliges, O., Hook, J., Löchtefeld, M., Motamedi, N., Muller, L., Olivier, P., Roth, T., von Zadow, U.: Multi-touch surfaces: A technical guide. Tech. rep., University of Munich (2008)
25. Schöning, J., Hecht, B., Raubal, M., Krüger, A., Marsh, M., Rohs, M.: Improving Interaction with Virtual Globes through Spatial Thinking: Helping users Ask "Why?". In: IUI '08: Proceedings of the 13th annual ACM conference on Intelligent User Interfaces. ACM, USA (2008)
26. Schöning, J., Krüger, A., Cheverst, K., Rohs, M., Löchtefeld, M., Taher, F.: Photomap: using spontaneously taken images of public maps for pedestrian navigation tasks on mobile devices.

In: Proceedings of the 11th International Conference on Human-Computer Interaction with Mobile Devices and Services. pp. 14:1–14:10. MobileHCI '09, ACM, New York, NY, USA (2009)
27. UNIGIS. Guidelines for Best Practice in User Interface for GIS: ESPRIT/ESSI project no. 21580 (1998)
28. Usländer, T. (ed.): Reference Model for ORCHESTRA Architecture V2, Project Deliverable. Orchestra Consortium (2007)
29. Vogel, D., Balakrishnan, R.: Interactive public ambient displays: transitioning from implicit to explicit, public to personal, interaction with multiple users. In: UIST '04: Proceedings of the 17th annual ACM symposium on User interface software and technology. pp. 137–146. ACM (2004)
30. Wasinger, R., Stahl, C., Krüger, A.: M3I in a Pedestrian Navigation & Exploration System. In: MobileHCI '03: Proceedings of the 5th International Symposium on Human-Computer Interaction With Mobile Devices and Services. Springer (2003)
31. Weiser, A., Neis, P., Zipf, A.: Orchestrierung von OGC Web Diensten im Katastrophenmanagement – am Beispiel eines Emergency Route Service auf Basis der OpenLS Spezifikation. In: GIS – Zeitschrift für Geoinformatik. pp. 35–41 (2006)
32. Weiser, M.: The computer for the 21st century. Scientific American 3(3), 94–104 (1991)
33. Weiser, M., Gold, R., Brown, J.S.: The origins of ubiquitous computing research at PARC in the late 1980s. IBM Syst. J. 38, 693–696 (1999)
34. Weiser, M.: The computer for the 21st century. In: Baecker, R.M., Grudin, J., Buxton, W.A.S., Greenberg, S. (eds.) Human-computer interaction, pp. 933–940. Morgan Kaufmann (1995)
35. Wilson, A.D., Izadi, S., Hilliges, O., Garcia-Mendoza, A., Kirk, D.: Bringing physics to the surface. In: UIST '08: Proceedings of the 21st annual ACM symposium on User interface software and technology. pp. 67–76. ACM (2008)
36. Wobbrock, J.O., Morris, M.R., Wilson, A.D.: User-defined gestures for surface computing. In: CHI '09: Proceedings of the 27th international conference on Human factors in computing systems. pp. 1083–1092. ACM (2009)
37. Wu, M., Balakrishnan, R.: Multi-finger and whole hand gestural interaction techniques for multi-user tabletop displays. UIST '03: Proceedings of the 16th annual ACM symposium on User interface software and technology pp. 193–202 (2003)
38. Wu, M., Shen, C., Ryall, K., Forlines, C., Balakrishnan, R.: Gesture Registration, Relaxation, and Reuse for Multi-Point Direct-Touch Surfaces. In: TABLETOP '06: Proceedings of the First IEEE International Workshop on Horizontal Interactive Human-Computer Systems. pp. 185–192. IEEE (2006)

Perspectives on Reasoning About Time[1]

Martin Charles Golumbic

Abstract Reasoning and acting within the time constraints of the real world are among the most fundamental notions of intelligence. Understanding the nature and structure of such constraints can help to find a satisfying solution or find a relaxation when no solution can be found. Given certain explicit temporal relationships between events, we may have the ability to infer additional relationships which are implicit in those given. For example, the transitivity of "before" and "contains" may allow inferring information regarding the sequence of events. Such inferences are essential in story understanding, planning and causal reasoning. Temporal information may be qualitative where events are represented by abstract time points and time intervals, and we process and deduce relationships between them, such as pairs intersecting each other, one preceding, following or containing another, etc. Other information may be quantitative where durations can be measured, precise time stamps may be available, or numerical methods can be applied to understand a specific time line of events. We will explore a variety of these topics.

1 Introduction

The spirit of this morning's lecture is to exercise our brains on some higher-level mathematics.

$$\mathbf{2009 - 20 = 1989}$$

---

[1]This chapter is based on a talk given at the German–Israeli Minerva School in autumn 2009 at the University of Haifa in Israel.

M.C. Golumbic (✉)
University of Haifa, Haifa, Israel
e-mail: golumbic@cs.haifa.ac.il

A. Krüger and T. Kuflik (eds.), *Ubiquitous Display Environments*,
Cognitive Technologies, DOI 10.1007/978-3-642-27663-7_4,
© Springer-Verlag Berlin Heidelberg 2012

Fig. 1 Temporal reasoning

Where were we 20 years ago? Most of the students in this lecture were in elementary school, yet this was about the time when I started looking seriously at the topic of temporal reasoning. Today every school pupil has a mobile phone, and every university student a laptop; back then there were few laptops and no cell phones. Nobody ever thought seriously about building the current types of embedded systems which my colleagues have been discussing in earlier lectures. Perhaps they were "thinking" about it in 1989, but many of the important devices did not yet exist.

2009 + 20 = 2029

Where will you be in 20 years? Perhaps you will be high-tech managers, professors, owners of companies; maybe you will be living on a farm or something similar if high-tech is not for you. Many of your current professors, like me, will be retired or almost retired and the systems that we will need in 20 years, when some of us will barely walk with a cane, are all of the systems that you have been talking about in this workshop. Your professors are relying on all of you to provide *them* with the technology that they are going to need when they are 80 years old! I don't know what it will be, but that is what *you* are going to have to figure out.

This paper will explore something more theoretical in the hope that, using your creativity and imagination, you will see how it might relate to the topics of this workshop).

2 The Dinosaurs

Temporal Reasoning is an old science that goes back very far, at least as far back as the paleontologist in the picture who is looking at these bones. He is wondering, "What did the dinosaur really look like?" He has some partial information with which he is attempting to reconstruct what actually may have existed – how it appeared, and what really happened (Fig. 1).

As computer scientists, when you analyze the logs of a computational process and monitor someone's motion over 24 h – possibly using the mobile phone to which they reluctantly subscribe – you are going to have to look at what may seem to be too much information. Or, maybe you will take samples and find that you do not have enough information. Either way, you will need to extract and decipher what the real, hidden information is all about. That is what you will be interested in.

In temporal reasoning, something similar occurs: you may have some information about what has happened in the last 20 min, or the past 20 years, and you need to reconstruct what actually happened from the clues that you are given, because you cannot really have all of the history.

The areas of reasoning about time events in which I have been interested, and have worked over these past 20 years are the methods that have to do with constraint-based problems (CSP) – scheduling, planning and analyzing (problems and their resources). There will be a set of events that occur over time which you may get as partial data from some logs, or from another source. The figure below (Fig. 2) shows an example of scheduling university lectures during certain time periods, where we want to assign lectures to rooms. This is a fairly standard graph theory problem, one of coloring an intersection graph [7]. The time intervals correspond to the vertices of a graph, known as an *interval graph*, where two vertices will be joined by an edge in the graph if their intervals intersect (conflict in time). This very special case of graph coloring can be carried out very efficiently (unlike its more general cousin, the NP-complete coloring of arbitrary graphs).

Returning to the paleontologist in our picture, he may be reasoning about some less structured kind of time intervals, like the period in which a certain dinosaur lived, by studying elements of the bones of dinosaurs. He takes note that certain "Dinosaur S emerged before T and perished not after it". This is an example of the sort of temporal information the paleontologist may have about some prehistoric events or objects from long ago, and he may be able to deduce temporal relations between them. From such sets of interval relations we derive conclusions about what actually may have happened over time.

Building any sort of system, and looking at it from a time perspective, will raise issues that require considering the granularity of the temporal data. For example, do you really have a "second by second" log of where this person was, and what she was doing all of the time, or do you have some other, much less detailed, temporal information? You might rather look at what is important and decide from that perspective what the granularity should be. For example, when monitoring a real-time system for a nuclear power plant, what happens from second to second may be

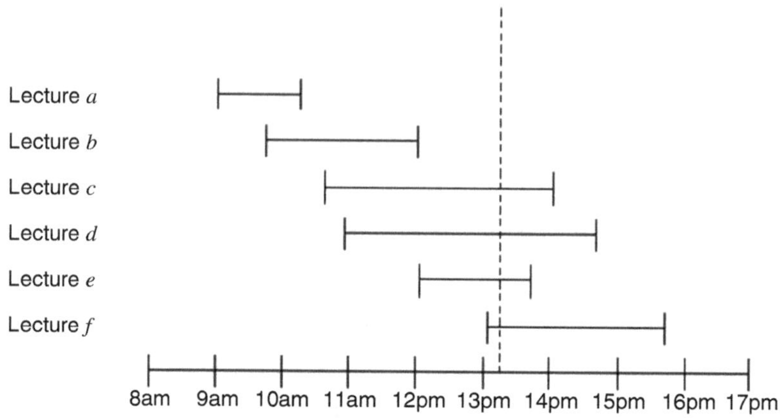

Fig. 2 Interval graph example

of great interest. However, if you reason about what someone did this morning at home, it is not really of much importance that he stepped from this spot, to that spot, to another spot; rather, you will be interested in a higher level set of activities, e.g., she got up, went to the kitchen, took her pills, (or forgot to take her pills), then she spent 32 min walking on the treadmill. The granularity may be in terms of nanoseconds, minutes, days, or epochs. If you're reasoning about those dinosaurs, then the day-to-day or year-to-year aspects are unimportant.

There is another important distinction to be made when designing a system that uses temporal information, namely, whether one regards time as "time points" or "time intervals". Even here the granularity aspect comes into play; for example, a physicist may be very interested in the duration of all events of an explosion seeing them as intervals each measured in nanoseconds. An engineer planning a set of charges to go off while building a tunnel under the Carmel mountain will probably just care about the time points at which one charge and another charge are set off, and what the implication will be of the order in which they are detonated.

Then there is the further issue of persistence of the world, like whether certain things cause particular events to persist, or take time, or last forever. Whenever dealing with any kind of temporal system, in fact, one has to deal with change due to actions, partial information, and synchronizing time lines. When one person's account of some event and another person's account of the same event have to be somehow merged, where are the contradictions? Where do they actually agree or mesh? Can they be put together to tell a real story?

What Time Is It?

Every Monday morning for years, at about 11:30 am, the telephone operator in a small Nevada town received a call from a man asking the exact time.

One day the operator summoned the nerve to ask him why the regularity.

"I'm a foreman of the local sawmill," he explained. "Every day, I have to blow the whistle at Noon. So I call you to get the exact time."

The operator giggled, "That's really funny," she said. "All this time, we've been setting our clock by your whistle."

3 A Little Help from My Friends

To be fair, although I have presented some of my own perspective, I have also asked some of my friends for theirs:

What does ***Zeno*** say about time?

The Paradox of Achilles and the Tortoise is usually read in philosophy courses. "In a race, the quickest runner cannot overtake the slowest." It is important to think about this when designing a real-time system. The Arrow Paradox, if you follow this logic then: "You cannot even move."

Aristotle disputed Zeno's reasoning.

Time is not composed of "nows". If there is just a collection of "nows" then there is no such thing as temporal magnitude. If there is just a collection of "nows" then there is no notion of duration.

Heraclitus goes with the flow.

Change is the only constant in the Universe. "On those who step in the same river, different and different waters flow." The only thing that we can count on as being constant is that it will always change.

Keith Cheverst's thoughtful lecture in this workshop[2] raised some interesting points:

A message is left at a smart door panel, "Back in 15 minutes", raises questions:

- How can it be temporally updated? Should there be a count-down?
- How was it used over time – analyze the logs? How do you analyze the logs?

These messages are of a qualitative temporal nature:

- "Out to Lunch" – what does this mean?
- "Back soon"
- "Working at home tomorrow"

Expectations – reliability of messages, things are written/posted on the door that are false.

We have certain expectations of what temporal knowledge is all about, and a lot of background knowledge is needed. What do we have to put into the system? How do we do it? One often guesses wrong about what the users really think about.

What do some of my current friends consider to be the most important accomplishments and challenges in Temporal Reasoning?

---

[2] Cheverst, Taher and Fisher chapter "The Design, Deployment and Evaluation of Situated Display-Based Systems to Support Coordination and Community' in this book."

Michael Fisher sent this list of issues (Handbook of Temporal Reasoning in Artificial Intelligence):

- Model checking
- Temporal reasoning to XML querying
- Alternating-time
- Temporal aspects of natural language
- Exploring the limit of decidability in first-order temporal logics

Angelo Montanari thought that formal specification and verification in reactive systems was the most important thing that temporal reasoning was doing.

> **Alfonso Gerevini:** Perhaps the most important accomplishments in the last 20 years are Allen's interval algebra (IA), Vilain & Kautz's point algebra (PA), and the Golumbic & Shamir classes. (Thanks very much!)

3.1 The Berge Mystery Story

At some point you may wish to read the *Berge Mystery Story,* a kind of temporal reasoning story that I have always found quite interesting, it is written by Claude Berge, a well known French mathematician and good friend. We won't solve his problem today, but you can find it in Ref. [5] or on the Internet.

4 Allen's Temporal Interval Algebra

One model of temporal reasoning is Allen's Temporal Interval Algebra A_{13}. I will try to expose you to the "hooks" into the subject, which many of you have already seen in your AI courses, or maybe not. Nonetheless, you need to know that they exist in the literature – and that people actually use temporal reasoning systems.

Allen's Temporal Interval Algebra A_{13} looks at all the ways that two intervals x and y could be related – they can be moved around in 13 different ways, as Table 1 shows (the interval y is bold, the interval x is light.) The bottom pair consists of two equal intervals, the second pair from the bottom has the property that they do not start at the same place, but x finishes y, or equivalently, y is finished by x, etc. Reasoning about the possible relationships that exist between intervals provides another tool that can be incorporated into a system that has to deal with time events.

4.1 Qualitative Temporal Reasoning of Events

Allen's Temporal Interval Algebra is an example of a qualitative model, one that has these properties:

- No mention of numbers, clock times, dates, etc.
- Relations such as *before*, *during*, *after* or *not after* between pairs of events.

Table 1 Allen's temporal interval algebra A_{13}

Relation	Notation	Interpretation
x before *y*	$\prec$	
y after *x*	$\succ$	
x meets *y*	m	
y met-by *x*	m^{-1}	
x overlaps *y*	o	
y overlapped-by *x*	o^{-1}	
x starts *y*	s	
y started-by *x*	s^{-1}	
x during *y*	d	
y includes *x*	d^{-1}	
x finishes *y*	f	
y finished-by *x*	f^{-1}	
x equals *y*	$\equiv$	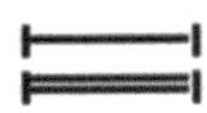

- Algorithms that are used to process information through propagation of constraints and constraint satisfiability between pairs of events using backtrack search.

Audience Question
Does Microsoft Project use intervals?
Answer
I have no idea. I would imagine that all actual planning systems have to incorporate not only qualitative but mostly quantitative reasoning using intervals.

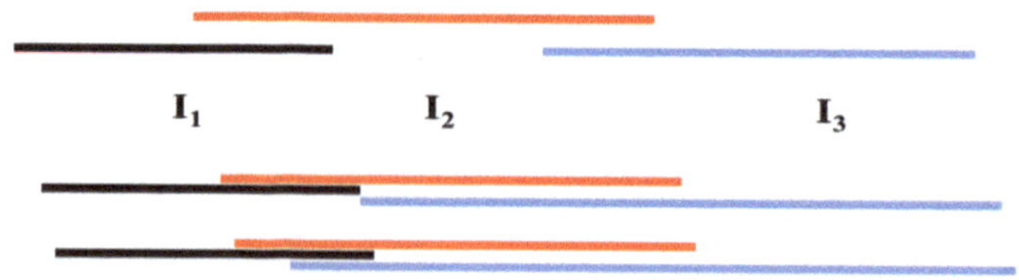

Fig. 3 Temporal interval algebra example

We'll see a model later that uses quantitative information. For example, the meaning of 9 o'clock is quantitative; there is no such measurable meaning for "after" without saying something like, "23 minutes after".

As an example of the kind of reasoning that Allen does in his model, consider the following. You have three intervals; you know that event 1 *meets* event 2, and event 2 *meets* event 3. From that you can deduce that event 1 happened sometime *before* 3, and there is a time gap between them. You can see that just by drawing a diagram, but formally can you incorporate this reasoning by the rule

$$\text{if } I_1\, m I_2 \text{ and } I_2\, m I_3 \text{ then } I_1 < I_3$$

This is a kind of extension or generalization of the familiar notion of transitivity.

Here is a slightly more complicated example:

You know that 1 and 2 *overlap* and that 2 and 3 *overlap*. Could several configurations satisfy this? If so, what would those be? If I_1 o I_2 and I_2 o I_3, it could look like any of these (Fig. 3):

Looking again at Table 1, the relationship between 1 and 3 can be one of three possibilities, namely: I_1 $\{<, m, o\}$ I_3. In other words, I_1 is strictly *before* I_3, or they *meet*, or it could be that they *overlap*. Several possible scenarios – a *disjunction* of possibilities.

4.2 *Allen's Algebra as a CSP*

Allen's Algebra and the methods used by it can be regarded as a constraint satisfiability problem (CSP). Here we have:

- Variables $R_{i,j}$ for all pairs of intervals
- Each variable takes values from A_{13}
- Domains are disjunctions of the current possibilities
- Algebraic Closure:
  - Look-up table or calculation
  - Propagation of constraints

A text book by Rina Dechter on solving constraint satisfiability problems shows how to apply CSP techniques to solve temporal problems (see Ref. [2]).

4.3 *Coarser Algebras and Fragments*

Thirteen interval relations are a lot! That gives you $2^{13}-1$ non-trivial disjunctions to reason about. In our paper [6], Ron Shamir and I decided to consider some "macro" relations and reduce the number of basic relations: $\cap < >$. Defining intersection by

$$\cap = \{s,\ f,\ d,\ m,\ o,\ \equiv\ s^{-1},\ f^{-1},\ d^{-1},\ m^{-1},\ o^{-1}\}$$

we obtain the coarser algebra A_3

There are other coarser algebras which will not be discussed here in detail, called A_6 and A_7, but imagine that in the full interval algebra of Allen there are 13 possible relations between pairs, that is 8191 disjunctions, a big number to work with, whereas in the Golumbic-Shamir algebra there are $2^3-1=7$ disjunctions. Of course, there is a "coarseness" price to pay. This means that we are not going to reason about the endpoints of the intervals, whether they *overlap*, *meet*, *include*, etc.; we block out or ignore or never had endpoint knowledge. We lose information, but we have a more concise representation and can more easily deal with the algorithmics of reasoning in the coarser algebra A_3: $\cap < >$.

We offer here an illustrated example taken from Ref. [4].

Goldie and the Four Bears

Once upon a time there were four bears, Papa bear, Mama bear, Baby bear and Teddy bear.

Each bear sat at the table to eat his morning porridge, but since there were only two chairs, (the third chair was broken in a previous story), the bears had to take turns eating.

Baby and Teddy always ate together sharing the same chair, and on this day Mama was seated for part of Baby bear's meal when the door opened and in walked their new neighbor, Goldie.

"What a great aroma," Goldie said. "Can I join for a bowl?" Mama replied, "Sure, but you will have to wait for a chair!"

"Yeah, I know all about chairs," said Goldie. So Goldie sat down when Baby and Teddy got up.

Papa entered the kitchen. Looking at Mama, he said, "I wouldn't sit at the same table with that girl." Mama answered, "Then it's good you ate already."

In analyzing the temporal elements this story, as illustrated above in Fig. 4, one can ask a number of questions such as:

- Could Papa and Baby both be at the table together?
- Could Papa and Mama both be at the table together?
- Could Papa have spent some time at the table with both Baby and Mama?
- Did anyone sit at the table with Goldie?

Facts from the Story

- Only two chairs (spatial not temporal information.)

Fig. 4 Goldie and the Four Bears

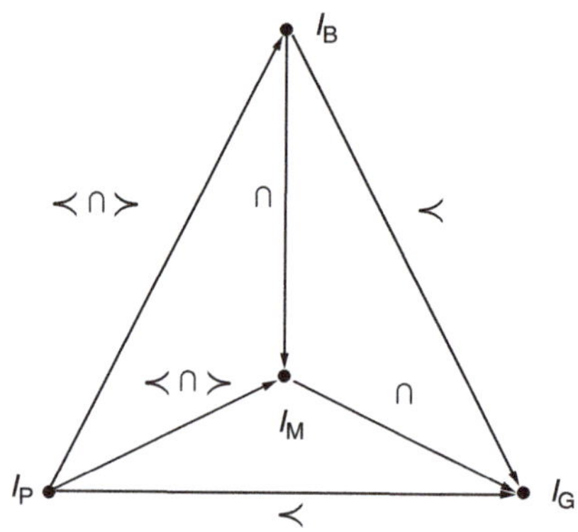

Fig. 5 The constraint graph

- $I_B \cap I_M$: Mama and Baby were seated when the door opened. (The interval for Baby and Mama is non-empty, since we know from the story that they were there at the same time.)
- $I_B < I_G$: Goldie sat down when Baby got up. (We know that the interval from when Papa and Baby ate is strictly less than the interval when Goldie ate.)
- $I_P < I_G$: Papa ate before Goldie.
- $I_M \cap I_G$: Papa to Mama (seeing her seated): (We know that the interval when Mama is eating and the interval when Goldie is eating intersect, because Papa said, "I wouldn't sit . . . with that girl.')

From this story, one can deduce additional temporal information. Formally, you have a *constraint graph* where the constraints/relations are put on the edges, gathered from the input above: the Allen relations between intervals I_B, I_G, I_P, I_M (Fig. 5).

Then, by using a constraint propagation algorithm, you try to reduce the number of possibilities. For example, the following rules act like a transitivity table (Fig. 6),

In this way, propagation deletes some impossibilities. It is *impossible* that Papa was at the table after Mama: $I_P < I_G$ and $I_G \cap I_M$, so by our rule, we may delete the relation $>$ on the edge from I_B to I_M.

Fig. 6 Constraint rules

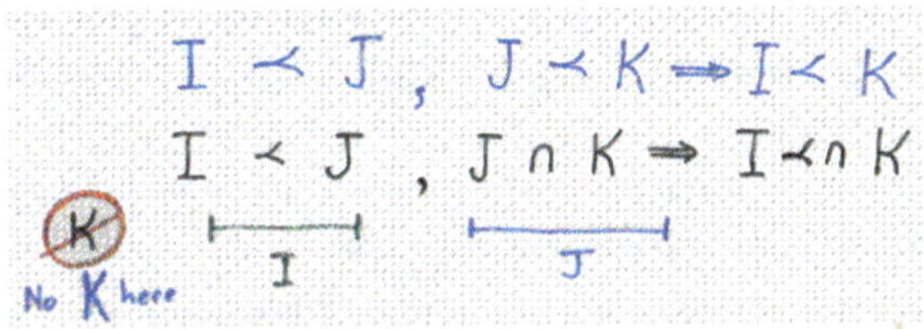

5 Complexity of Testing Interval Consistency

Allen's algebra A_{13} and the coarser algebra A_3 give rise to the following:

The Interval Satisfiability Problem (ISAT): Given a disjunction of interval relations for each of the variables $R_{i,j}$ (between pairs of intervals). Is there a collection of intervals that satisfy these constraints?

The computational complexity of the Interval Satisfiability Problem, and of the sub-problems defined by restricting the domains to its specific fragments, is one of the fundamental theoretical results in temporal reasoning. For the general case, the following hold:

Theorem 1. *(Golumbic and Shamir [6]) ISAT is NP-complete for A_3.*

Corollary 2. *(Vilain and Kautz [11]; with bug fixed by Vilain et al. [12]) ISAT is NP-complete for Allen's Algebra A_{13}.*

Thus, reasoning with just the three macro interval relations is already computationally hard, not to mention Allen's full set of 13. However, it turns out that not everyone uses 8191 different disjunctions when talking about the relationships between 100 intervals and 5 intervals. For example, the fragment consisting of only singletons in A_{13} is known to be polynomial. Also, in many types of real applications, by the nature of the application, not all possible relations will occur. In such cases, by considering just a fragment of the relations, some problems become tractable.

We now present the computational complexity of all 31 symmetric fragments[3] of A_3.

Theorem 3. *(Minimal Intractable Fragments) ISAT remains NP-complete even when we have only labels from* $\{\cap, <>, <\cap>\}$, [6], *or labels from* $\{<\cap, \cap>, <>\}$ [13].

Theorem 4. *(Maximal Tractable Fragments) ISAT is linear when we have labels only from* $\{<, >, \cap, <\cap, \cap>, <\cap>\}$ or from $\{<, >, <>, <\cap>\}$, *and is $O(n^3)$ when we have labels only from* $\{<, >, \cap, <>\}$, [6].

[3] In A_3 there are $2^3 - 1 = 7$ disjunctions of which two are symmetric pairs to two others, namely $<\cap$ with $\cap>$ and $<$ with $>$; thus, there are only five essentially different disjunctions; hence, $2^5 - 1 = 31$ symmetric fragments to be considered.

The implication of the first result in Theorem 4, for example, says that if you leave out the relation of disjointness $< >$ allowing only the remaining six relations, then you can test satisfiability in linear time. *What does that mean?* It means that the language *forbids* me to make a statement like, "Tsvika got to the office and left *either* before Eyal was there *or* after Eyal was there." This bad kind of uncertainty in the interval relation, namely either *before* or *after* without knowing which one, is precisely what raises the computational complexity so dramatically from linear to intractable. If you are so fortunate as not to have those kinds of relations, thus finding yourself in a tractable case, then you can use constraint propagation and solve the satisfiability question efficiently.

Why is this important? It is important if you are going to try and run an actual system that does temporal reasoning.

Finally, we should mention that, like the complexity studies of fragments of A_3 that we have just seen above, there has been a decade of work on tractable subalgebras of A_{13}, most notably first by Nebel and Bürckert [9] where they looked at the fragment called the ORD-Horn, a subalgebra of the disjunctions of A_{13} containing all the 13 basic relations and closed under converse, intersection and composition. ORD-Horn covers about 11% of A_{13} (but does not include $< >$). Then, in the period 1997–2003, in a series of papers by Christer Bäckström, Thomas Drakengren, Peter Jonsson, Andrei Krokhin and others, a comprehensive investigation took place on the tractability and intractability of fragments of A_{13}, see Ref. [8] and its references.

6 Point Algebras

A further, perhaps easier, related topic is the study of various *point algebras*, where we reason about the disjunctions of the point relations $\{<, =, >\}$. At first glance, this would seem to be very similar to A_3, however, intervals are much different from points, although the symbols of the relations may look the same. It turns out that complexity-wise they are very different.

Theorem 5. *Satisfiability with the point algebra is polynomial-time solvable in all cases.*

Dozens more papers have been published in the past 15 years on variations of the theme of combining the qualitative interval relations with the point relations, studying the computational complexity of their fragments.

7 Discussion

Audience Question
Are there systems that take into account time, web searches, and things like information being more valuable that stand the test of time.

Answer
Surely they do a very partial job.

Audience Question
You said that if there is time branching, either I go this way or that way, then the complexity increases, even though you've got more information about the situation. For example, if you start bottom-up with a real-world situation you get more information and then you say, well, I cannot compute it properly.

Answer
Sometimes knowing less is better.

Audience Question
But maybe you can make a random choice to say throw information out. If I have too much information, good statements, but one statement that is a bit tricky, get rid of that information.

Answer
In any kind of search where you have disjunction, you have alternatives, say a or b. You start branching your search, you're search space expands exponentially, and then you have to decide what to do. It doesn't matter whether it's a temporal domain or any other type of search procedure. You may ask, "Is the search space small, do you have enough of an estimation function to be able to prune it and do back-track search successfully?" Or is the space just too big and you can't, so you fail. Or do you just make some assumptions and say we accept the assumption that Tsvika came first, because he's always on time! Then based on that assumption, which may be false because the train was late today, I will continue in this scenario, and I might get to a false conclusion.

8 Simple Temporal Problems

In my AI seminars, I often like to talk about a more quantitative model for temporal reasoning, called simple temporal problems (STP) and temporal constraint satisfaction problems (TCSP). I refer you to an older paper by Dechter et al. [3] introducing the topic, my survey paper [4], and two recent papers by Choueiry and Xu [1, 11] where new efficient algorithms are presented. I will not go into detail, I'll just show you two examples.

Example (STP)
Tom has class at 8:00 a.m. Today, he gets up (P_1) between 7:30 and 7:40 a.m., and then prepares his breakfast (P_2) taking 10–15 min. After eating breakfast (another 5–10 min), he drives to school (20–30 min). Will he be on time to class?

Here is a similar kind of graphical model to those that we have seen, but where you have time points being the vertices of the graph and relations between one event and another event being directed edges of the graph. In this model, temporal events are represented only by time points, like in the (qualitative) point algebra, so an interval must now be represented by its two end points. There is an upper and

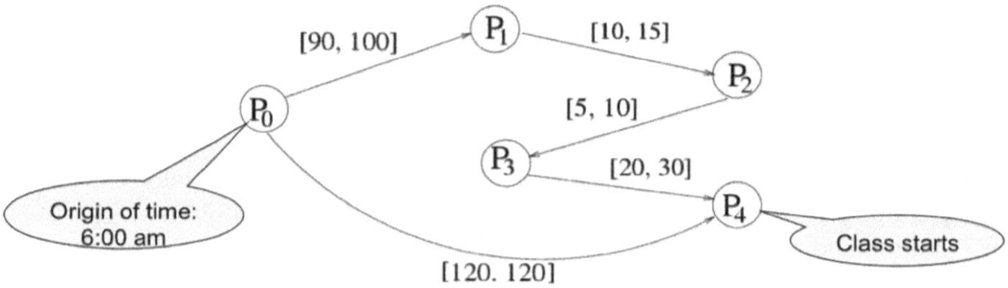

Fig. 7 Graphical representation of STP example

lower bound on each directed edge indicating the duration of moving from the first time point to the second time point. For example, in Fig. 7, the edge from P_1 to P_2 labeled [10, 15], may be interpreted as saying that the time taken to move between the event P_1 and the event P_2 is somewhere between 10 and 15 min.

If one starts with a story like in this example, with temporal information which is not exact but is numerical, you basically must ask, Is this story consistent? What's the earliest possible time that Tom could have gotten to school? And what's the latest possible time? Various different algorithms are used for solving such simple temporal problems, some more complex than others. One method is to use the familiar Floyd–Warshall algorithm from graph theory; others may use known techniques in AI such as those that are known as path-consistency.

This example looks somewhat similar to the constraint graph for Goldie and the Four Bears in Fig. 5, but with numerical labels. There are relations between three items (a path with two edges), and one can propagate the constraints to reduce the size of the labels on some of the edges. For example, the figure below taken from Ref. [14] shows part of an STP where Partial Path-Consistency allows us to reason as follows, updating the labels of the edges (i.e., tightening the constraints). First, the label on edge AC is reduced from Refs. [2, 12] to [8, 12] since the path from A via B to C indicates that it must take at least 8 min to get from A to C in a consistent solution; then BC can be updated from Refs. [2, 7] to [2, 6]. In this way, we have tightened the constraints.

Simple temporal problems can be solved by efficient polynomial time algorithms. The more general case of "not so simple" temporal problems, namely, temporal constraint satisfaction problems (TCSP), which allow *disjunctions* of intervals as labels, have much higher computational complexity. They are often solved (non-optimally) by heuristic algorithms (Fig. 8).

9 Another Application

I wanted to find something to speak about that was especially relevant to the topic of this workshop, and found one paper that I will now show you. I hope it may inspire you to do something really innovative in the future.

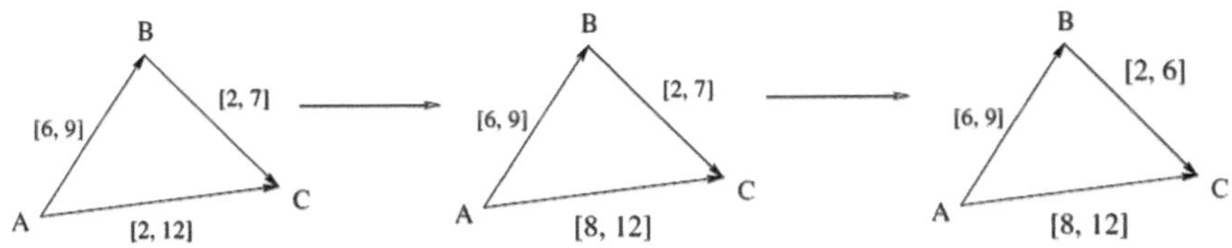

Fig. 8 TCSP example

Massimo and Tsvika are very interested in museums[4] – monitoring movement and assisting the visitor in a variety of automated ways. There was a paper at an AAAI workshop in 2008 called *Incorporating Temporal Reasoning into Activity Recognition for Smart Home Residents* [10] that actually hit closer to home. The key problem to be solved there is activity recognition in a home setting. In temporal reasoning, you are looking to identify temporal relations between events and discover intervals that are meaningful. The same would be true in temporal data mining. In this paper, the authors are interested in the application of *at-home health monitoring* (which, in 20 years from now, I too will be very interested, and hope *you* will have developed good systems).

In my favorite museum, I might like to visit once a week or once a month and have all sorts of embedded systems suggesting to me what I should do and see, based on what it knows I have done in the past, or what I might like. At home, I'd also very much like to be reminded about things by a friendly system: to take my medicine, to know something about my telephone use, to help me with my financial management, personal hygiene, and my hydration and food consumption. Such topics were proposed by caregivers at an old-age home where they were dealing with people with Alzheimer's. What are the needs that such a system should serve? I think the samples suggested here may be pretty mundane, but they leave a lot of room for the imagination. What systems should be incorporated into my monitored home of the future?

Here is my townhouse in an assisted-care facility in the year 2029 (Fig. 9). It's going to have lots of sensors and monitors, and I will wear an electronic necklace that is going to let the monitoring system know all sorts of things about what I do. Of course, it doesn't need to know everything that I do – you must filter the important things from the unimportant things. *That's the challenge*. When we look at what I do from minute to minute, on the one hand, we look at the top level activities, at what *kinds* of things I am doing, and need to anticipate what *kinds* of things I want done. But when we look at this environment as computer scientists and engineers, we must ask what has to be in the "guts" of the system? What are the algorithms that are going to have to actually be used in order to know whether I fell down, and whether I got up again?

[4] Kuflik, Boger and Zancanaro chapter "Analysis and Prediction of Museum Visitors' Behavioral Pattern Types" in this book.

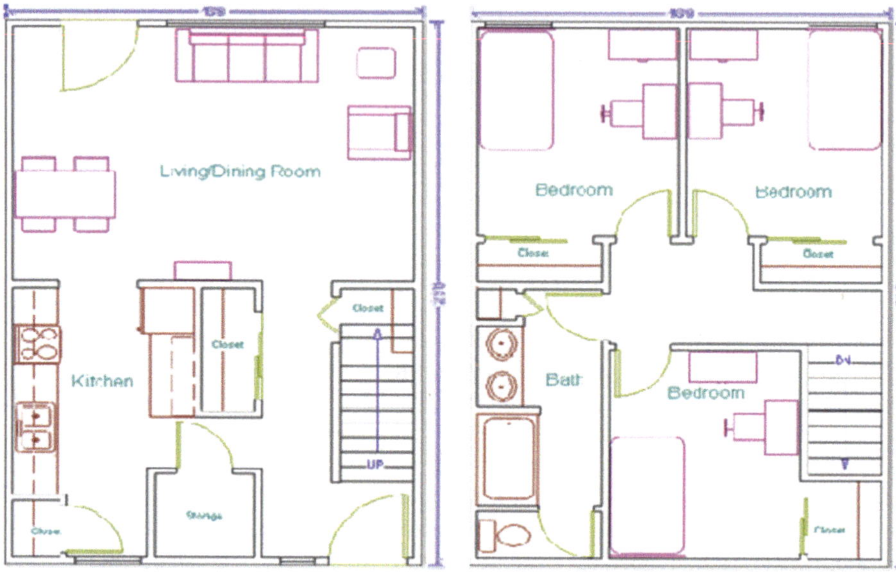

Fig. 9 My townhouse

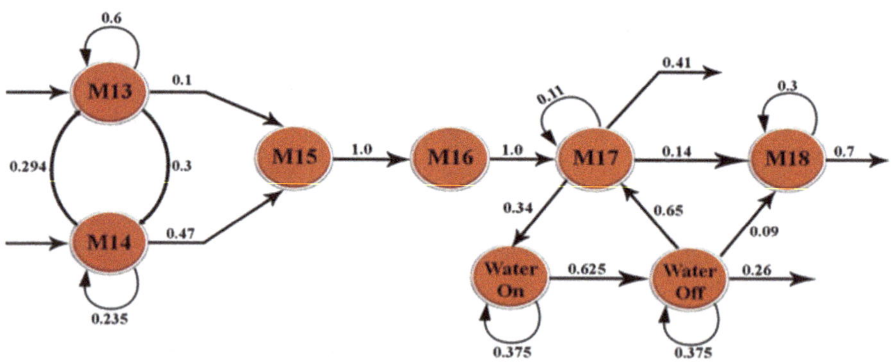

Fig. 10 Markov model example

The system advocated in the paper uses Markov model to recognize a few activities: washing hands, cooking, phone call and clean up. The diagrams above (Fig. 10) illustrate different stages of getting up and taking a drink of water: what are the probabilities that you go from one stage to the next? and at what times? It looks rather annotative. What's the actual distribution of the probabilities for the water staying off or turning on, moving from one stage to the other, and so on. It's a very nice exercise of what kind of models might be needed to handle certain monitoring and recognizing whether I'm getting up to get a drink of water or whether I'm getting up to take my medication.

One needs to carefully read the paper. What is important is raising the concepts, and asking whether their solution is likely to lead to the right solution. Perhaps, five of you graduate students here may be able to figure something out by next year, which will be much better. Then, one needs to carefully read *your* paper.

10 Claude Berge and Robert Aumann

Quoting Claude Berge, "The use of a mathematical tool may be unexpected." This really says something to me as a mathematician and computer scientist. We never know when the mathematical tools discovered 20 years earlier might be useful for something today.

In Sarit's talk,[5] she had a picture of Prof. Aumann from the Hebrew University who received a Nobel Prize several years ago. He gave a talk at the Technion earlier this year[6] at a special event with Nobel winners giving general lecture to the public. Prof. Aumann included a story – he was recently amazed to find that some theorem of *his*, that he proved when he was about 1 year past his doctoral thesis in New York, something on Braid Groups and group theory, a nice mathematical theory, had an unexpected application. He was talking with his grandson who is a medical student at Ben Gurion University, and lo and behold he finds out that in the medical school classes they were talking about braid groups. They were discussing how different blood vessels intertwine, yet not forming knots. And there was his grandfather's theorem. So you never know.

11 The End Is Just the Beginning

We have seen a few of the many possible temporal models that may be needed to take into consideration when building real user-oriented systems. I hope that I have made my point: almost every system that you will build has some element in time. From this perspective, for time oriented problems, one model is going to be more suitable than another, depending what your project is about. For the application, what is going to be useful for me, a Markov model, an interval reasoning model, point reasoning, and so on? You need to have that familiarity of what's out there, in order to put it into any system that you build. I'm counting on you for the year 2029.

[5] Lin and Kraus chapter "From Research and Practice: Automated Negotiations with People" in this book.

[6] Robert J. Aumann: "Pure Science and Applied Science". In: Rambam Maimonides Medical Journal 2(1):e0017 (2011).

References

1. Berthe Y. Choueiry and Lin Xu: An Efficient Consistency Algorithm for the Temporal Constraint Satisfaction Problem. In: AI Communications, vol. 17 (4), pp. 213–221. IOS Press (2004)
2. Rina Dechter: Constraint Processing. Morgan Kaufmann (2003)
3. Rina Dechter, Itay Meiri and Judea Pearl: Temporal constraint networks. In: Artificial Intelligence, vol. 49, pp. 61–95. Elsevier (1991)
4. Martin Charles Golumbic: Reasoning about time. In: Frederick Hoffman (ed.) Mathematical Aspects of Artificial Intelligence: Proc. Symposia in Applied Math, vol. 55, pp. 19–53 American Math. Society (1998)
5. Martin Charles Golumbic: Graph theoretic models for reasoning about time: In: Lecture Notes in Computer Science, vol. 3321, pp. 362–372. Springer (2004)
6. Martin Charles Golumbic and Ron Shamir: Complexity and algorithms for reasoning about time: a graph-theoretic approach. In: J. Assoc. Comput. Mach., vol. 40, pp. 1108–1133. ACM (1993)
7. Martin Charles Golumbic and Ann N. Trenk: Tolerance Graphs. Cambridge University Press (2004)
8. Andrei Krokhin, Peter Jeavons and Peter Jonsson: Reasoning about temporal relations: The tractable subalgebras of Allen's interval algebra. In: J. Assoc. Comput. Mach., vol. 50, pp. 591–640. ACM (2003)
9. Bernhard Nebel and Hans-Jürgen Bürckert: Reasoning about temporal relations: a maximal tractable subclass of Allen's interval algebra. In: J. Assoc. Comput. Mach., vol. 42, 43–66. ACM (2003)
10. Geetika Singla, Diane J. Cook, and Maureen Schmitter-Edgecombe: Incorporating temporal reasoning into activity recognition for smart home residents. In: Proc. of the AAAI Workshop on Spatial and Temporal Reasoning, pp. 53–61 (2008).
11. Marc B. Vilain and Henry A. Kautz: Constraint Propagation Algorithms for Temporal Reasoning. In: Proc. AAAI, pp. 377–382. (1986)
12. Marc B. Vilain, Henry A. Kautz and Peter van Beek: Constraint propagation algorithms for temporal reasoning: A revised report. In: Daniel S. Weld and Johan de Kleer (eds.) Readings in Qualitative Reasoning about Physical Systems, pp. 373–381. Morgan-Kaufman (1989)
13. Adam Brooks Webber: Proof of the interval satisfiability conjecture. In: Annals of Mathematics and Artificial Intelligence, vol. 15, pp. 231–238 (1995)
14. Lin Xu, Berthe Y. Choueiry: A new efficient algorithm for solving the simple temporal problem. In: 10th International Symposium on Temporal Representation and Reasoning/4th International Conference on Temporal Logic TIME, pp. 212–222 (2003)

Shared Interfaces for Co-located Interaction

Massimo Zancanaro

Abstract Co-located, collaborative work around shared surfaces has become a major topic of the research agenda in the fields of teamwork and collaborative learning. In this chapter, we introduce the difference between multi-touch and multi-user interfaces and we present three cases of co-located interactions on shared interfaces. The first two cases present examples of multi-user applications: StoryTable is a multi-user application for children to practise their narration skills in a collaborative way; the NNRT table is a multi-user application to foster a shift in the attitudes of participants via a narration task (it is used in peace-education programs). Finally, we describe a new approach to shared interfaces explicitly aimed at influencing immediate behavior in an informal, non goal-oriented co-located small group.

1 Introduction

Co-located, collaborative work around shared surfaces has become a major topic on the research agenda in the fields of teamwork and collaborative learning. Shared surfaces are computer systems based on large interactive surfaces placed horizontally (in this configuration they are usually called tabletop devices) or vertically (often called wall displays) on which a specifically designed interface is displayed or projected. They represent a radical shift from the paradigm of one user/one computer. As such, they are subject to design constraints that are different from those of standard graphical user interface (GUI) applications.

Relevant distinctions among devices that support shared surfaces are whether they support single or multi-access and single or multi-users (see Fig. 1). The distinction single/multi-users refers to the capability of the device to recognize

M. Zancanaro (✉)
FBK, via Sommarive 18, Povo 38050, Italy
e-mail: zancana@fbk.eu

A. Krüger and T. Kuflik (eds.), *Ubiquitous Display Environments*,
Cognitive Technologies, DOI 10.1007/978-3-642-27663-7_5,
© Springer-Verlag Berlin Heidelberg 2012

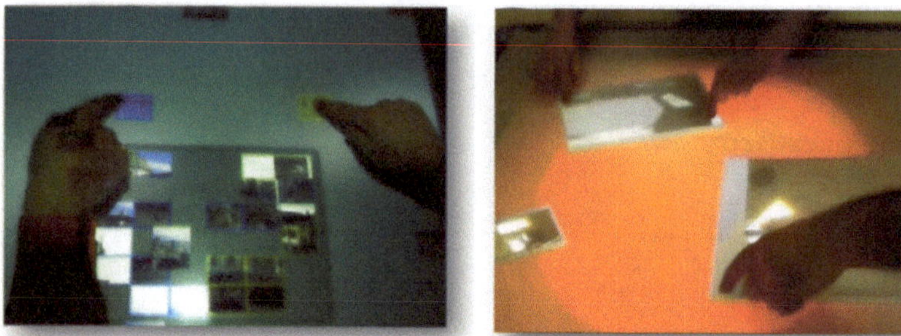

Fig. 1 Multi-touch (*left*) and multi-user (*right*) devices

different users when they interact with the system, that is, to track who is touching where. The distinction single/multi-access refers to the capability of the system to allow multiple touches on the surface, independently tracking each single touch; it thus allows complex gestures and simultaneous usage (but without recognizing the different users).

An example of the former is DiamondTouch developed by MERL [1] and now commercialized by the Boston-based company CircleTwelve.

Regarding multi-touch devices, the most well-known is MS Surface, a Microsoft commercial product that has the form factor of a small coffee table. MS Surface is based on a multiple-camera proprietary solution for recognizing an unlimited number of simultaneous touches on the surface while a computer display is back-projected on it. Other techniques for recognizing multi-touches are currently being investigated in research laboratories around the world. For example, there is a public domain technique based on a single camera approach, called FTIR (Frustrated Total Internal Reflection) whose specifications were made available by Han [2]. It is worth noting that several hardware manufacturers are now introducing to the market relatively cheap devices for multi-touch.

Shared interfaces support the collaborative use of resources and, at least potentially, foster group activities. In particular, using multi-user devices, some constraints may be placed on the interaction as part of the design of the interface in order to promote or scaffold collaboration.

This chapter summarizes three experiences with interfaces for shared surfaces that were explicitly designed to foster collaboration.

2 Multi-User Devices and Enforcing Collaboration

Collaboration in a strict sense means working jointly with others to pursue a common goal. In an educational setting, the framework of Collaborative Learning posits that students working in small groups tend to learn more and in a more effective way than students working alone [3]. Technologies may provide an

Fig. 2 Conceptualization of a multi-user touch gesture

effective support for collaboration provided that they are designed properly. In this chapter, we will explore a few ways of constraining the interaction on a tabletop device in order to foster a collaborative activity.

Morris et al. [4] discuss the notion of *cooperative gestures* where the system interprets the gestures of more than one user as contributing to a single, combined command. They present and evaluate different purposes of these types of gestures.

By exploiting DiamondTouch's capability to distinguish between simultaneous input from multiple users, we extended the event system of standard touchable interfaces (that is, the touch, the double touch, and the drag-and-drop events), allowing multiple-user events such as the multiple-user touch, double touch, and drag-and-drop. For instance, a multiple-user touch is an event that is triggered when two or more users click together on a button (Fig. 2).

Since we wanted the new events to extend the standard event system, we have a chain of events generated by user actions. In the case of two users touching a button, the system first triggers a standard touch event when the first user touches the button; when a second user touches the same button, a new multiple-user touch event is triggered. When one of the two users releases the touch, a multiple-user release event is triggered. Finally, when the other user releases the button the standard release event is triggered.

The multiple-user drag-and-drop event is a little bit more complicated because it requires that users not only touch the objects together (and, therefore, trigger the chain of events described above for the multiple-user touch), but also that they move their fingers on the surface in a synchronized way. Since this can be difficult to achieve, a threshold for avoiding an unexpected closing of the event was determined empirically.

3 Experience 1: StoryTable

Storytelling is a learning activity that allows children to develop skills such as creative problem-solving, collaborative learning, expressive design, the development of multiple forms of literacy, and the exploration of knowledge [5].

Recently, there has been growing interest in the use of novel technologies that support collaborative interaction among children. Among many, KidPad and the Klump [6] and the Magic Carpet [7] were developed as collaborative storytelling technologies for young children, subtly encouraging collaboration so that children can discover the added benefits of working together (as an example, children were able to combine digital crayons to produce new colours).

Also, there has been some exploration of technologies that allow children to reflect after performing activities outside the classroom: for example, Ambient Wood [8], where children first explored a digitally augmented woodland to find out about plant life in the different habitats, and then, back at the classroom, used an interactive screen display to show their information to one another and start forming predictions. Similarly, the Savannah project is a collaborative, location-based game in which groups of children role-play being lions on a hunt in a virtual savannah; an interface for replaying a system-recording of each completed level is then provided in the classroom so the children can reflect on how they performed [9].

The use of physical tools has also been explored to support storytelling in systems such as MIT's KidsRoom [10], StoryRooms [11] and StoryMat [12]. Storytent, a system developed at Nottingham University, is a projected display shaped like a tent which allows children to immerse themselves in a virtual world [13]. The TellTale interface [14] explores the idea of providing ways of making the pieces of a story "tangible". TellTale is a caterpillar-like toy with five modular, coloured body pieces on which children can record their own voices; once linked together, the audio is played back in sequence.

Some research has been focused on having the system work as a companion of a child. Sam, a tool developed at Media Lab [15], consists of a 3D, animated virtual child that can pass physical toys back and forth to real children and can cooperate with children in constructing stories.

3.1 Design Concept

The design was inspired by a methodology known as Family Bears [16], where a child is invited to play with puppets representing a family of bears and their living environments (the house, the school, and so on) and then to invent stories about what happens to the family. This approach is used in both educational and therapeutic settings to measure the linguistic capabilities of children and their relations with the world.

Cooperative Learning [17] is an instructional method that involves students working in teams to achieve a common goal. The emphasis is on the cooperation between the children that must be established and appropriately maintained in order for the methodology to demonstrate its benefits. To this end, a number of main tenets must be met. First of all, a *Positive interdependence* must be established among students, so that they feel linked with group mates in such a way that they cannot succeed unless their group mates do and they feel that they must coordinate their own efforts with the efforts of their group mates to accomplish the task. Positive interdependence has different facets; *Positive Goal Interdependence*, when the individual goals equal those of the group and *Positive Resource*

Fig. 3 The interface of the StoryTable on the MERL's DiamondTouch

Interdependence, when the resources needed to solve the problem have to be shared by the group members. Another condition is face-to-face *Promotive Action*, where individuals facilitate each other's efforts to achieve and complete task objectives in order to reach the group's goals. Finally, the performance of individual students must be assessed and the results provided to the individual and the group, fostering *Individual Accountability/Personal Responsibility*.

StoryTable is a multi-user application for DiamondTouch for two children (although DiamondTouch can work on up to six users, in this case the interface has been designed for only two because that is the optimal situation for collaborative learning).

Several scenarios are provided each of which has different backgrounds and characters for which children can invent stories. The interface displays some ladybugs wandering around the table surface (Fig. 3). A mixture of standard touch events and the new multiple-user events were used as a means to control the objects.

One ladybug carries the backgrounds, that is, the stages on which the story will be set, for example a wood or a medieval castle. In order to set the stages, the ladybug is opened by double touching on it and the chosen stage is dragged out. Since the selection of the background is crucial for determining the story, the children need to agree on which one to choose, and the system fosters the agreement by requiring that the drag is jointly performed by the children. This design choice is meant to apply the *face-to-face promotive actions* principle of Collaborative Learning.

Another ladybug carries the various characters and elements such as the Princess, the Knight, and so on. Each character can be dragged onto the current background (the size of the table allows each child to reach all the characters on the surface easily). Again, the ladybug can be opened by a single-user double touch event. However, in this case, the elements can be dragged autonomously by each child.

A third type of ladybug of a different shape (the blue ones in Fig. 3) contains the audio snippets that will form the story. In order to record some audio, a child has to drag one of these ladybugs into the recorder and then keep the button pressed while speaking. The audio snippets are recorded independently by the two children and—according to the criterion of *individual accountability*—once loaded with audio the ladybug displays a coloured shape meant to represent the child who recorded it.

Once loaded, an audio ladybug can be modified by the child who recorded it, but the system refuses modifications attempted by the other child.

In some sense, the ladybug is "owned" by the child who records it. Yet, the two children may agree to free a ladybug by a multiple-user drag-and-drop action. That is, if they drag the ladybug together onto the recording tool, the system removes the content and the highlighting. The ladybug is now free to be used by anyone. Again, this design choice is meant to foster face-to-face promotive actions.

Following the principle of *limited resources,* just six Audio Ladybugs are available following the criterion of positive resources interdependences and no more than 30 s of voice can be recorded on each ladybug. The system does not force the children to share the ladybugs, but we expect that the indication of the child who recorded each one—by means of the coloured shape—is enough to foster an appropriate use of collaborative skills and group processing.

To summarize, the main objectives pursued with the design of the interface are related to the basic principles of cooperative learning. The story is composed collaboratively by both children (positive goal interdependence) using a limited number of Audio Ladybugs (positive resources interdependence); each contribution is clearly attributable to the child who recorded it (individual accountability). Each child is allowed to work on an individual level, but crucial operations have to be performed by both children simultaneously (appropriate use of collaborative skills and group processing).

The limited portion of audio that each Ladybug may hold encourages children to reflect on the surface structure and the words they use to form their stories. Moreover, the possibility of reorganizing Audio Ladybugs will promote reflection on the coherence among story segments. Finally, by splitting the narration among different Audio Ladybugs, children are encouraged to consider the role of each story unit in their narrative.

3.2 Assessment

An evaluation study was conducted involving 70 children from two primary schools. The children were paired according to the results of a sociometric test so as to have neutral pairs (that is, children that have neither a strong positive nor a strong negative attitude toward their partner). The experiment was a between-subject study with a puppet-based (no tech) storytelling activity as the control condition. The two conditions were randomized. For this study, we had three hypothesis, namely: (1) StoryTable induces more complex and mature language because the children have the possibility to listen to the content recorded in the audio ladybugs and reason critically about it; (2) StoryTable does not induce less structured and less cohesive stories than the control condition; (3) the contribution to the story and the interaction between the children are more balanced in the StoryTable than in the control condition.

In the first phase (Fig. 4a), the children were separated in two rooms and invited to tell a story individually about using puppets. This initial phase was meant to

Fig. 4 The three phases of the experiment: an individual child narration (**a**), a pair of children using StoryTable (**b**) and the control condition using puppets (**c**)

assess the literacy ability of the children. The story produced was analysed considering the number of words and propositions, the number and structure of episodes, and the global cohesion (the coding scheme proposed for the Family Bears task; Bornstein *unpublished* and adjusted for Italian by Cesari and Venuti [18] was used).

Then, the two children were invited to tell a story together using StoryTable (Fig. 4b). This phase was meant as a training session for the system and the stories produced were not analyzed.

Finally, the children were invited to tell a story using StoryTable and, after a short break, to tell another story using the puppets (Fig. 4c). The two tasks were randomized. The two stories produced were then analyzed as discussed above.

A multivariate analysis with repeated measures showed an overall significant within-subject effect (Wilk's lambda $= 0.19$, $F = 203.792$; $p < 0.01$, observed power $= 1$). Univariate tests showed significant differences between the number of propositions ($p < 0.01$), the number of episodes ($p < 0.01$), and the number of words produced during the interaction ($p < 0.01$).

Consistently with the first two hypotheses the children's stories had significantly more propositions and episodes in the control condition than with StoryTable; at the same time the two conditions did not significantly affect either the quality of the stories produced, which were substantially identical both in terms of narrative structure and cohesion, or the number of words that made up the stories.

For the third hypothesis, a more balanced contribution to the story and the interaction, a multivariate analysis with repeated measures ($p < 0.05$) confirms that there is an effect due to the conditions, but a significant difference can only be found for the interaction ($p < 0.01$) while the difference in the story's contributions did not reach significance.

The lack of differences in the story contribution may be due to the weakness of the index used: the test may not be sensitive enough given the small sample. A qualitative assessment of these differences may be provided by analysing the plots of the Lorenz curves obtained, considering for each pair the story as a measure shared between the two children. In the case of perfect equality in the contribution, the graphical representation is a straight line of 45°.

Figures 5 and 6 represent the plots of the curves for the words used in the story in the control condition and the ST condition, respectively. It can be noted that

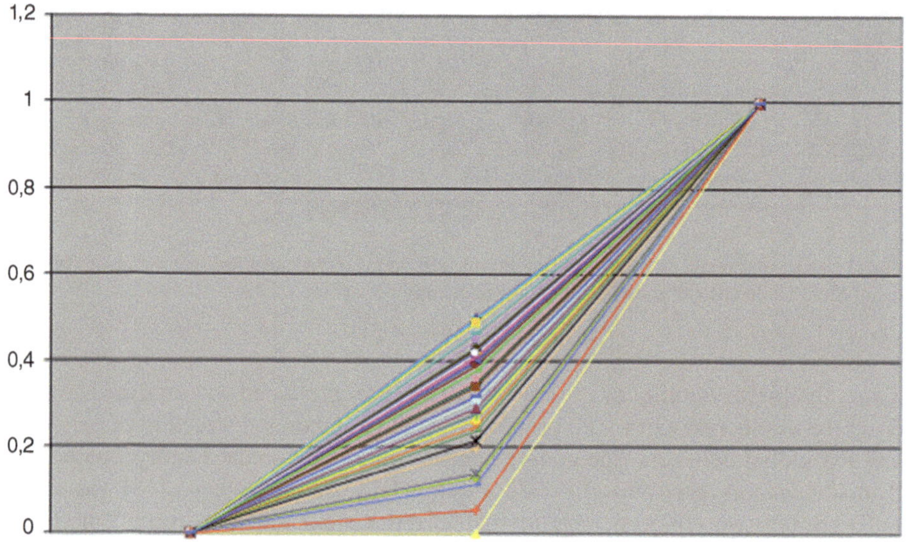

Fig. 5 Lorenz curve for the BT condition for the number of words composing the story: the x-axis represents the two children and the y-axis the distribution of the contributions to the narration

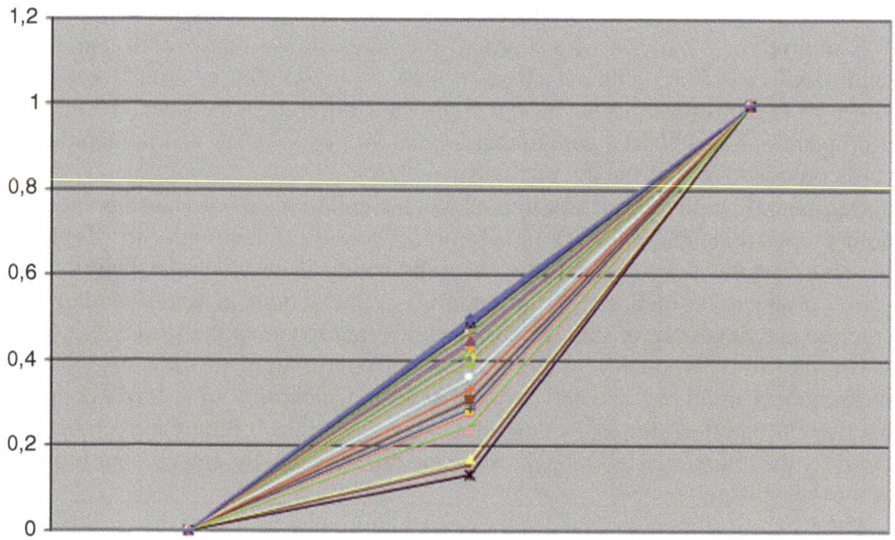

Fig. 6 Lorenz curve for the ST condition for the number of words composing the story: the x-axis represents the two children and the y-axis the distribution of the contributions to the narration

the curves for the ST condition are closer to the line of the perfect equality than the curves for the BT condition. Similar graphs are obtained for the other variables.

In conclusion, the use of shared resources on the table seemed to be effective in fostering a more mature language in the narrations. The extent to which the system actually fosters collaboration, in the sense of more balanced narrations, could not be fully assessed but the evidence in this sense is promising.

More details on StoryTable can be found in the book chapter [19]. The system has also been used as a tool for an educational intervention for children with autism; the experience was described in Ref. [20].

4 Experience 2: The Narration and Negotiation Reconciliation Table

In another prototype, we explored the role of technology in fostering a shift in the attitudes of participants via a narration task. The general claim is that participants may achieve a greater understanding of and appreciation for the other's viewpoint under conditions that support partaking in creating a shared narration. The work is based on two cultural elements: social psychology approaches to conflict resolution that emphasize the role of narration in small group interventions and the design of a collaborative tabletop interface aimed specifically for the task. In particular, we were interested in how a co-located interface can augment a typical face-to-face moderated discussion.

Several concrete initiatives have used narration as a methodology for improving the recognition of the other. For example, the Coexistence Workshop is a course in the teacher education program of the University of Haifa's Faculty of Education [21]. Bar-On and Adwan [22] engaged a group of teachers in a process of producing a joint textbook that includes two narratives with Palestinian and Jewish teachers. Maoz et al. [23] describe the use of narration in a mixed group of teenagers in Israel.

For our experimentation, we were particularly interested in the conflict between Palestinians and Jews, a conflict over the same territory that has continued for over a century in various forms. Our own work started from the recognition of narration as a key element for an educational initiative targeted at Jewish and Palestinian youth co-existence, as already discussed by Bar-On and Kassem [24]. The main characteristic of our work is the use of a specifically designed co-located shared interface that supports two teenagers in the production of a joint narrative.

The design of the interface was based on the idea of providing a setting for face-to-face multimedia narration with contributions from both sides and a joint, mutually acceptable outcome. The key point of the interface was to emphasize explicitly specific points of disagreement, rather than leaving them to be perceived as vague feelings of conflict and detachment, and to provide a means for acting jointly toward achieving a narration acceptable to both viewpoints. The process had to implicitly include classical steps in conflict resolution approaches, such as escalation and de-escalation [25]. Similarly to StoryTable, the requirement of physically acting together in the process of revising and completing the narration is added in the belief that tangible joint actions are an important component for achieving common goals.

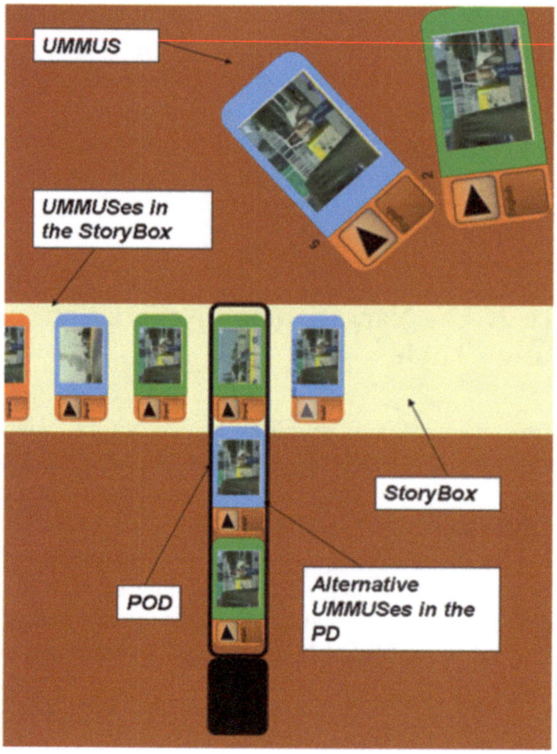

Fig. 7 A detail of a screen shot of the NNR-Table with the main components

4.1 Design Concept

The Narration Negotiation and Reconciliation (NNR) Table is a multi-user application for DiamondTouch. The participants sit facing each other on opposite sides of the table and a graphical interface is top-projected onto the DiamondTouch device (see Fig. 7). With this configuration, the participants have different points of view of the interface. Although this may create problems for the visualization of the elements (see for example Kruger et al. [26]) as well as being more confrontational, we thought that a shoulder to shoulder positioning on the same side of the table would not be appropriate for this paradigm.

The basic element of the interface is the UMMUS (Unit of MultiMedia User Story) which also sounds similar to a typical middle-eastern food much liked by both populations in Israel. An UMMUS can be loaded with a picture and 20 s of voice can be recorded. An UMMUS can be recorded with audio in the participant's own language as well as its English translation. Each UMMUS has a color that represents the participant who created it (colors that have national meaning for each side—green for the Arab Palestinian participant and blue for the Jewish participant).

A sequence of UMMUSes creates the story. In order to add an UMMUS into the sequence, a participant just drags and drops it into a container (called the StoryBox) positioned in the center of the interface. In contrast, removal of an UMMUS from the story requires cooperative gestures: the two participants need to drag it out together. An initial pre-defined UMMUS is automatically positioned as the first element of the story in order to give the participants an initial narrated segment. The StoryBox is a place where the participants can negotiate the process of "sequencing" some of the UMMUSes produced in order to create a coherent shared narration.

The basic design constraint is that any constructive action (e.g., creating an UMMUS or placing it in the StoryBox) may be performed by each participant independently while any removal action (e.g., removing an UMMUS from the StoryBox or deleting an UMMUS) has to be done through a joint action.

A crucial element of the interface is the possibility of expressing and negotiating disagreement in a tangible way. A *Point of Disagreement (POD)* is a frame that can be positioned around an UMMUS in the StoryBox and has the effect of "labelling" that contribution as an object of disagreement. Once a POD has been positioned, the participants cannot conclude the story (the "Done" button is automatically disabled). In this sense, PODs induce "lose-lose" situations. At the same time PODs are tools for proposing alternatives: one or more additional UMMUSes representing alternative formulations out of the *impasse* are dragged into the frame.

A POD can be removed when the two participants agree upon which UMMUS they want to keep as an acceptable formulation. They do so by jointly pressing the POD removal button. Again, a cooperative gesture is used to mark their awareness of an important event (in this case, the positive result of a negotiation). Additional UMMUSes and PODs may be added in an iterative manner but the story will remain incomplete until all PODs have been resolved.

4.2 Assessment

A qualitative study was conducted in Haifa consisting of nine pilot interventions. All of them involved two Israeli males, a Jew and an Arab, aged 17–18 years. Each intervention included a joint narration task using the NNR-Table (carried out in Hebrew) and two interviews (one pre-task and one post-task) carried out respectively in Hebrew and in Arabic for the two participants (Fig. 8).

The purpose of the pre-task interview was to document attitudes towards the conflict in the region and toward the other participant as a member who is stereotypically considered to be the enemy. The purpose of the post-task interview was twofold: (1) to document a possible change in attitude towards the conflict and toward the other participant and (2) to evaluate the intervention including the role of the technology in enhancing a possible change in attitude toward the conflict and toward the other participant.

The material used as a stimulus included a set of photos that represented a range of both positive and negative situations in the life of young Israelis. Among other

Fig. 8 Interaction with the NNR-Table

things, these photos depicted an Arab family having dinner together, Jewish and Arab youngsters on a beach, as well as negative situations such as a Jewish mother crying at a funeral and Israeli soldiers searching a Palestinian house.

A trained moderator was present during all the sessions. The role of the moderator was mainly aimed at keeping the subjects focused on the narration task and encouraging them to use the NNR-Table functions.

Although we could not observe a clear-cut change in participant attitudes, all expressed general satisfaction and sometimes even surprise at learning more about the partner's viewpoints. In some cases, the participants were explicitly made aware that their stereotyped opinions of the "other" side were incorrect. This appears to be related to the exchange of story ideas that was facilitated by the audio-visual media and turn-taking paradigm.

One aspect that was made clear is that it is crucial to select and pair the subjects in such a way as to have an escalation of conflict. It is more difficult to obtain results if participants are politically too extremist or too moderate.

The central aspect of the design, the POD, proved useful for encouraging the creation of alternate points of view in non-threatening ways. Yet for some participants, the POD appeared to be a somewhat strong tool for signaling disagreement. On the other hand, it was useful in escalating the conflict and this is a basic aspect of our approach. In the post-intervention interview, one participant gave a nice interpretation of the POD, saying that it encouraged a cultural discussion because if one disagrees one does not have to interrupt the other, but present his view later; everything is kept in the software.

In general, the novel cooperative and co-located setting and the availability of multimedia elements appear to be interesting and motivational for the participants.

Participants commented positively on the uniqueness of engaging in an activity-based interaction—"doing"—and not just talking.

More details on NNR-Table can be found in Refs. [27] and [28].

5 Experience 3: The Augmented Café Table

In the previous two experiences, we described two examples of multi-user applications for shared co-located interaction. Both applications require the direct interaction and focused attention of their users. In this section, we introduce a new type of system explicitly aimed at influencing immediate behavior in an informal, non goal-oriented co-located small group. The state of the group dynamics is assessed in order for the system to plan and deploy minimalist strategies continuously using evocative means to influence behavior rather than explicit recommendations. A key aspect of our approach is that the main "interaction channel" is left for direct human-to-human interaction, while no large conscious elaboration effort nor actions are meant by the user toward the interface.

This approach has some affinity with peripheral displays [29] in that the interface is not central to the attention of the group and people may look at it only occasionally. Yet peripheral displays are meant to have a passive role, for they usually display easily graspable information (such as weather or stock graphics), aiming at making the user just aware of it without it becoming the central focus of attention. Our system takes a different stance, aiming at staying at the periphery most of the time while attracting the users' attention when contextually more appropriate.

The specific prototype presented here is a tabletop device that looks like a "normal" café table. On its surface the system is able to display subtle visual cues that aim to encourage the conversation to move to a certain topic or to sustain the conversation on the current one. The reason for investigating this scenario is that recent studies [30] recognize the importance of conversations within small groups of visitors as a way of improving the museum experience and the quality of learning. We believe systems of this kind are potentially very useful in the educational domain, especially where informal learning environments are concerned.

5.1 Design Concept

Tired after the visit to an art museum, four friends stop at the museum café for a drink (Fig. 9a). They start chatting and they notice that, on the surface of the table they are sitting around, some fish in a Zen pond are displayed while sometimes images from the museum exhibits appear in the water (Fig. 9b). Peter is bored as Paul and Mary talk about work that is waiting them the next day .

Peter notices that the image appearing "in the water" in front of him, where the fish converge, is about something that attracted his interest during the visit and mentions this to Paul and Mary. Slowly the other two also intervene and Mary sees something moving in her area, with a detail zoomed in after the image has rotated

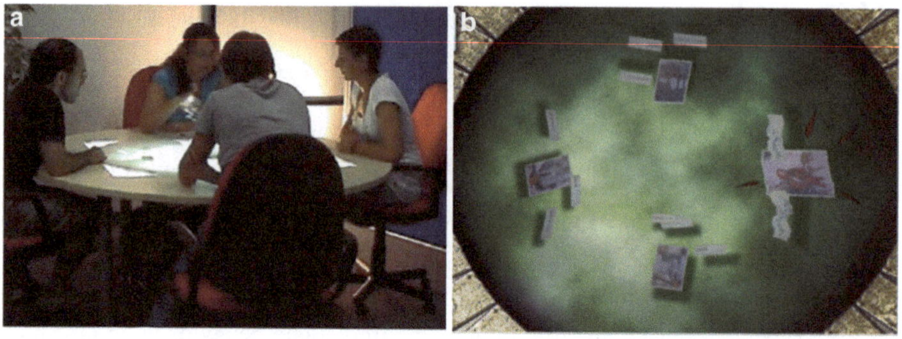

Fig. 9 The system in use (**a**) and a screenshot of the tabletop display (**b**)

around Peter's image; she senses that this is clearly related to the ongoing theme of discussion so she comments. While the table keeps displaying floating images and gently swimming fish, the three of them enjoy a good conversation about what they have just seen in the museum.

The system is based on a graphical interface top-projected on a standard table around which people can sit, enjoy a drink and talk. While participating in the conversation, people may be attracted by images or text purposefully displayed on the table. The display is not interactive in a strict sense, since users are neither requested nor allowed to manipulate the objects displayed on the table. Presentation generation is based on the system's perception and representation of the group behavior and driven by appropriate communicative strategies.

The perception modules process data from the visual and the acoustical scene. Cameras and microphones are hidden at the center of the table. Two kinds of information are used: a face detector processes the video streams from four cameras and estimates the visual attention of each individual participant; a voice activity detector processes the audio from four microphones positioned on the table in order to perceive who is contributing to the possible conversation.

The system attempts to affect the group behavior indirectly by presenting contextually appropriate visual material in a novel way, reminiscent of the tradition of advertisements: attention catching, evocative, and cognitively stimulating. Animation of images and words is a key characteristics of the stimuli displayed on the table. Motion captures attention and is easier to identify in the periphery than color and shape [31]. Proper timing of animations is indeed of paramount importance since the onset of motion is more effective at capturing attention than motion itself [32].

The presentation strategies are triggered by contextual conditions and by communicative goals inspired by the strategies used in the world of advertising.

5.2 *Assessment*

The system's design was inspired by an initial *Wizard of Oz* study described in Ref. [33].

A second user study was conducted with an initial prototype implementing a simple detection mechanism and a few simple presentation strategies.

Three spontaneous groups of three subjects (students of the local university) were involved, and each was observed in a single session. Members of the same groups were already acquaintances.

Each group was invited to our lab to visit a reconstruction of the "Torre Aquila" frescoes in the Buonconsiglio museum in Trento. Subjects were given a four-page booklet to help them during the visit; they were told that the purpose of the study was to test the content of the booklet. After the visit people were conducted to another room and were invited to sit at a table while waiting for the experimenter to come back. The system started soon after the subjects were settled.

At the end of each session, an experimenter debriefed the group about the real purpose of the study and conducted a semi-structured interview aimed at eliciting participants' feelings.

In general, the table triggered interest and the subjects exploited stimuli as expected, to keep alive or resume the conversation. The cameras, though quite visible, were rarely noticed and they did not seem to have affected the group behavior in any relevant way. As expected, the more technically skilled people were more enticed by the technology itself than by its content. For the subjects of the other two groups, their interest in the system itself was not prevalent.

This study confirmed the important assumption that whenever a conversation was going on and one of the subjects kept on looking at the table, this was a reliable signal that he/she was not involved in the conversation. Indeed, in 30 of the 37 cases where a single subject kept on looking at the table he or she was not speaking at the same time.

The effectiveness of the system in persuading people to intervene in the conversation clearly emerged from the analysis of recorded data: in 24 of the 30 instances of individuals looking at the table without being involved in the conversation, that individual actually intervened immediately following stimulus presentation; moreover, in 12 the instances his/her intervention was clearly related to the stimulus.

After a stimulus displayed on the table has become the topic of the conversation, the table is mostly able to support the conversation for a while. Of the 79 observed events in which people were all looking at the table, the subjects were actually speaking about content related to stimuli on the table in 66 cases. We noticed also that the subjects tended to deictically refer to the presentation by pointing on the table. Usually, the subjects did not mean to interact with it but to refer to it in the conversation.

In several of these cases, the stimulus that originated the conversation was displayed at the center of attention for the whole group and the system then applied the strategy of introducing a new stimulus at the center of the table. This effect was quite distracting for those holding the conversation.

More cohesion is needed in the transition from one state to another, as discussed above, including a reasonable analysis of time. For example, entering the state in which all are looking at the table from the state in which just one was looking at it is likely to indicate that the suggestion was effective and therefore the same topic or another closely related should be maintained. Conversely, entering in the same state from the state in which nobody looks at the table is likely to indicate that the previous conversation theme was exhausted, so a new stimulus is probably necessary.

More information on the Augmented Café table may be found in Ref. [34].

6 Conclusion

In this chapter, we introduced two experiences of multi-user applications in which the constraints on joint performance of gestures were explicitly designed to foster cooperation. In both cases, we discuss user studies aimed at assessing the effectiveness of this approach. In both cases, we understood how a specifically designed system is able to foster collaboration by appropriately constraining the interaction.

We then described a new type of system explicitly aimed at influencing immediate behavior in an informal, non goal-oriented co-located small group. The state of the group dynamics is assessed in order for the system to plan and deploy minimalist strategies continuously using evocative means to influence behavior, rather than explicit recommendations. This system is still a lab prototype for examining the difficulties of integrating state-of-the-art perceptual components with a robust interface. Yet, as the robustness of the perceptual components increases this challenging field of research will become a more realistic one to pursue. Open questions that deserve further investigation include how to evaluate this kind of system and how to define more effective strategies.

Acknowledgements The work presented in this chapter is a summary of several projects worked together with many colleagues at FBK-irst (formerly ITC-irst) and the University of Haifa. In particular: Oliviero Stock, Fabio Pianesi, Daniel Tomasini, and Patrice L. (Tamar) Weiss.

References

1. Dietz, P., and Leigh, D.: 2001. DiamondTouch: A Multi-User Touch Technology. In: Proceedings of UIST 2001, 219–226.
2. Han, J.Y.: 2005. Low Cost Multi-Touch Sensing through Frustrated Total Internal Reflection. Symposium on User Interface Software and Technology: Proceedings of the 18th annual ACM symposium on User interface software and technology. Seattle, WA, USA, 115–118.
3. Johnson, D., and Johnson, R.: 1994. Learning together and alone, cooperative, competitive, and individualistic learning. Prentice-Hall, Needham Heights, MA.
4. Morris Ringel, M., Huang, A., Paepcke, A., and Winograd, T.: 2006. Cooperative Gestures: Multi-User Gestural Interactions for Co-located Groupware. In: Proceedings of ACM CHI-06. Montréal, Canada.
5. Peterson C., and McCabe A.: 1983. Developmental Psycholinguistics: Three Ways of Looking at a Child's Narrative. Plenum, New York.
6. Benford, S., Bederson, B. B., Åkesson, K., Bayon, V., Druin, A., Hansson, P., Hourcade J. P., Ingram, R., Neale, H., O'Malley, C., Simsarian, K. T., Stanton, D., Sundblad, Y., and Taxén, G.: 2000. Designing Storytelling Technologies to Encourage Collaboration Between Young Children. In: Proceedings of ACM Conference on Computer Human Interaction, CHI2000.

7. Stanton, D., Bayon, V., Neale, H., Ghalli, A., Benford, S., Cobb, S., Ingram, R., Wilson, J., Pridmore, T., and O'Malley, C.: 2001. Classroom Collaboration in the Design of Tangible Interfaces for Storytelling. In: Proceedings of Human Factors in Computing Systems (CHI 2001) ACM Press, pp. 482–489.
8. Price, S., Rogers, Y., Stanton, D., and Smith, H.: 2003. A new conceptual framework for CSCL: supporting diverse forms of reflection through multiple interactions. In: Wasson, B., Ludvigsen, S., Hoppe, U. (eds). Designing for Change in Networked Learning Environments. Proceedings of the International Conference on Computer Supported Collaborative Learning.
9. Benford, S., Rowland, D., Flintham, M., Drozd, A., Hull, R., Reid, J., Morrison, J., and Facer, K.: 2005. Life on the edge: supporting collaboration in location-based experiences. In Proceedings of the SIGCHI conference on Human factors in computing systems (CHI'05).
10. Bobick, A. et al.: The KidsRoom. Communications of the ACM 43: 3, pp. 60–61 (2000)
11. Alborzi, H., Druin, A., Montemayor, J., Platner, M., Porteous, J., Sherman, L., Boltman, A., Taxén, G., Best, J., Hammer, J., Kruskal, A., Lal, A., Schwenn, T. P., Sumida, L., Wagner, R., and Hendler, J.: 2000. Designing StoryRooms: interactive storytelling spaces for children. In: Proceedings of the 3rd Conference on Designing Interactive Systems: Processes, Practices, Methods, and Techniques, New York City, New York, United States (August 17–19, 2000).
12. Ryokai, K., and Cassell, J.: 1999. StoryMat: a play space for collaborative storytelling. In: CHI'99 Extended Abstracts on Human Factors in Computing Systems (Pittsburgh, Pennsylvania. (May 15–20, 1999).
13. Green, J., Schnädelbach, H., Koleva, B., Benford, S., Pridmore, T., and Medina, K.: 2002. Camping in the Digital wilderness: tents and flashlights as interfaces to virtual worlds, In: Proc. ACM CHI 2002 Conference Abstracts, 780–781, ACM Press.
14. Ananny, M.: 2002. Supporting children's collaborative authoring: Practicing written literacy while composing oral texts. Computer-Supported Collaborative Learning, Boulder, Colorado, January, 2002. Pp. 595–596.
15. Ryokai, K., Vaucelle C., and Cassell, J.: 2003. Virtual peers as partners in storytelling and literacy learning. Journal of Computer Assisted Learning. 19(2):195–208.
16. Bornstein, M. H. (ed): The Bear Family. *Cognitive Coding Handbook*. Unpublished Manual. National Institute of Child Health and Human Development. Bethesda (MD).
17. Johnson, D., and Johnson, R: 1999. Learning Together and Alone: Cooperative, Competitive, and Individualistic Learning. Allyn and Bacon, Boston.
18. Cesari, P., and Venuti, P.: 1999. The projective bears family. Presented at 16th International Congress on Rorschach. Amsterdam, July 1999.
19. Zancanaro, M., Pianesi, F., Stock, O., Venuti P., Cappelletti, A., Iandolo, G., Prete, M., and Rossi, F.: 2007. Children in the museum: an environment for collaborative storytelling. In: Stock, O., and Zancanaro, M. (eds.) *PEACH: Intelligent Interfaces for Museum Visits*. Cognitive Technologies Series, Springer, Berlin.
20. Gal, E., Bauminger, N., Goren-Bar, D., Pianesi, F., Stock, O., Zancanaro, M., and Weiss T.: 2009. Enhancing social communication of children with high functioning autism through a co-located interface. *AI and Society, the Journal of Human-Centred Systems*. Volume 24, Issue 1.
21. Luwisch, F.E.: 2001 Understanding what goes on in the heart and the mind: learning about diversity and co-existence through storytelling. Teaching and Teacher Education 17:133–146.
22. Bar-On, D., and Adwan, S.: 2006 The PRIME Shared History Project. Peace-Building Project Under Fire. In: Iram, Y. (ed.) Educating Toward a Culture of Peace, pp 309–323, Information Age Publishing, Charlotte, NC (2006).
23. Maoz, I., Bar-On, D., and Fakhereldeen, M.: 2002. The Dialogue between the "Self" and the "Other": a Process Analysis of Palestinian Jewish Encounters in Israel. Human Relations, 55(8): 931–962.
24. Bar-On, D., and Kassem, F.: 2004. Storytelling as a way to work-through intractable conflicts: The German Jewish experience and its relevance to the Palestinian-Israeli context. Journal of Social Issues 60(2):289–306.
25. Eisikovits, Z., and Buchbinder, E.: 2000. Locked in a violent embrace. Sage, Thousand Oaks, CA.

26. Kruger, R., Carpendale, S., Scott, S., and Tang, A.: 2005. Fluid integration of rotation and translation. In: Proceedings of ACM CHI-05. Portland.
27. Stock, O., Zancanaro, M., Rocchi, C., Tomasini, D., Koren, C., Eisikovits, Z., Goren Bar, D., Weiss, P.L.: 2008. A CoLocated Interface for Narration to Support Reconciliation in a Conflict: Initial Results from Jewish and Palestinian Youth. In: Proceedings of ACM CHI 2008. April, 2008, Florence, Italy (Honourable Mention, Best of CHI Awards).
28. Stock, O., Zancanaro, M., Rocchi, C., Tomasini, D., Koren, C., Eisikovits, Z., Goren Bar, D., Weiss, T.: 2009. The Design of a Collaborative Interface for Narration to Support Reconciliation in a Conflict. AI and Society, the Journal of Human-Centred Systems. Volume 24, Issue1.
29. Weiser, M., and Brown, J. S.: 1996. Designing Calm Technology, PowerGrid Journal, v1.01.
30. Leinhardt, G., and Knutson, K.: 2004. Listening in on Museum Conversations. Altamira Press.
31. Bartram, L., Ware, C., and Calvert, T.: 2005. Moticons: Detec-tion, distraction and task, International Journal Human-Computer Studies, 58 (5). pp. 515–545.
32. Abrams, R. and Christ, S. E.: 2006. Motion onset captures attention: A rejoinder to Franconeri and Simons (2005)', Perception and Psychophysics, 68 (1), pp.114–117.
33. Rocchi, C., Tomasini, D., Stock, O., and Zancanaro, M.: 2008. Fostering conversation after the museum visit: a WOZ study for a shared interface. In: Proceedings of the Conference on Advanced Visual Interfaces, AVI-2008, Napoli, Italy.
34. Stock, O., and Zancanaro, M.: 2009. Formative Evaluation of a Tabletop Display Meant to Orient Casual Conversation. In Know Techn Pol 22:17–23.

Considering the Aesthetics of Ubiquitous Displays

Noam Tractinsky and Eleanor Eytam

Abstract We call attention to the important role that aesthetics, or visual beauty, can and should play in the deployment of ubiquitous displays. Following a short survey of the emergence of visual aesthetics in the field of human-computer interaction (HCI), we suggest that the importance of aesthetics stem from three main perspectives. The design perspective suggests that beauty is a cornerstone of any design discipline and that it is often positively related to other practices of good HCI design, such as usability. From a psychological perspective we argue that aesthetics induces pleasure and satisfies human needs; and that it is influential in how people form first impressions and subsequent attitudes. Finally, we argue that aesthetics has an important practical role as a differentiating factor between interactive products. Even more importantly, it is already woven into the socio-technical web of our time. We discuss the implications of this position for the design of ubiquitous displays.

1 Introduction

Current research in the paradigm of ubiquitous displays or its variants (e.g. situated public displays or context adaptive digital signage) appears to concentrate on issues that relate to content (e.g. [45, 46]), functionality and control. To some extent, a discussion of usability issues has started emerging regarding the usage of this technology. In this paper, we would like to point attention towards another aspect – aesthetics, which we consider immensely important for the technology, for designers of ubiquitous display applications, for researchers in the field, and of course for the users and the intended audience of situated displays. We suggest that, like in many walks of life, especially those which include a significant visual

N. Tractinsky (✉) • E. Eytam
Ben-Gurion University of the Negev, Beer Sheva, 84105, Israel
e-mail: noamt@bgu.ac.il

A. Krüger and T. Kuflik (eds.), *Ubiquitous Display Environments*, Cognitive Technologies, DOI 10.1007/978-3-642-27663-7_6,
© Springer-Verlag Berlin Heidelberg 2012

component of information and message delivery, the aesthetics of situated displays will have an important role in their successful deployment.

The objective of this paper is to underline the importance of addressing visual design aspects of ubiquitous displays from several perspectives. We describe each perspective from the more general viewpoint of interactive technologies in general, and relate how each of those perspectives is also relevant to ubiquitous displays.

2 Background

Today's technology supports almost every activity in our every day routine. Still, additional large scale changes are on their way into our daily lives. One of these changes is likely to be in the mold of what Weiser termed ubiquitous computing (ubicomp). Ubicomp encompasses many aspects. Most pertinent to our chapter is an aspect often overlooked by current writing on ubicomp. We refer, in fact, to the central concept of ubicomp, which was termed by Weiser and colleagues [76, 77] "encalming" technology. This is a vision of technology that allows us to easily move "from center to periphery and back". It enhances our peripheral reach and our connectedness to the environment around us. At the roots of the search for encalming technology stands traditional technology, which is described by Weiser and Seely Brown as the "enemy of calm". While Weiser et al. do not specifically mention the aesthetic aspects of their envisioned technology, it is clear that their plea for the reduction of visual and auditory noise and to improve our perception of the environment's wholeness has strong relationships to aesthetic conventions (see also Ref. [39]).

In addition to the ubiquity of technology, the ubicomp vision also include the disappearance of computers, which are to be embedded in everyday objects such as wall-sized displays and tabletop computers situated in the environments and buildings we inhabit [1]. Together, these hallmarks of ubicomp create new and important aesthetic challenges to the computing community. These challenges join those experienced by human-computer interaction (HCI) researchers concerning the role of visual aesthetics in various realms of interactive computing – from ATMs to media players to Web sites. Below we briefly survey the history of HCI research that led to this research.

2.1 Some Historical Background

During the last decade we have witnessed a considerable shift in the mindset of researchers and designers of interactive technologies. From an industry concerned with functional and performance issues, the industry has shifted towards emphasizing design, form, and style [58]. Similar, though perhaps less dramatic developments took place in the academic community, where studies about the nature and impact of aesthetic design have started to emerge [15, 21, 23, 27, 68]. The increased interest in aesthetics among the industrial and academic communities may reflect the maturation of the HCI field and the overcoming of many of its growing pains as a discipline that struggles with unreliable technology on the one hand and with the need to satisfy users' bare necessities on the other hand. A more detailed account of this process is provided in Tractinsky [71, 72]. In this chapter we only outline how aesthetic research in HCI has emerged from its ancestor disciplines.

One may ask, why was aesthetic not present in HCI from the beginning? The reason can be seen in its evolution. The roots of the field of HCI can be traced to the turn of the twentieth century with the advent of the scientific management movement. The trigger for looking for ways to optimize man and machine interaction was the industrial revolution, which introduced mass production. The idea was that scientific management methods could remedy "inefficiency in almost all of our daily acts" [67]. Scientific management was mostly concerned with the planning and execution of mechanical, physical tasks, such as handling pig iron. Scientific methods, according to Taylor, could have quadrupled the workmen's productivity while improving their well-being [67].

Later on, with the beginning of the Second World War the concern over the interaction of men and machine grew more important due to the large scale introduction of new, relatively complex machines, for use by masses of soldiers with very little background and training. It soon became obvious that human errors in operating the new systems caused many casualties. The insight that developing complex systems have significant implications for those who operate them and for society at large was a catalyst for the new discipline of Human Factors (or HF as it is known in the United States) or Ergonomics (the corresponding name in Europe). While many ideas and methods were borrowed by HF from Scientific Management, a major development was the emphasis on the cognitive aspects of the task.

Over the years, information technologies have penetrated almost every aspect of human life. During the 1960s and 1970s time sharing systems were developed and computers were introduced in growing pace into the workplace. That was a major first step towards interactive computing, which was significantly augmented later during the 1980s with the proliferation of the personal computer. The widespread use of computers in everyday lives introduced further challenges that required additional specialization by the human factors community, leading to the emergence of the field of human-computer interaction. To a large extent, the basic tenets that drove design guidelines under the scientific management and human factors paradigms remained intact. However, due to the characteristics of computerized systems (which were of limited physical variety), the emphasis of HCI research and design moved increasingly towards cognitive consideration at the expense of the physical considerations. Still, the basic striving for efficient performance that drove scientific management and much of HF research could still be seen at the center of HCI work until very recently. For example, usability expert Jakob Nielsen demonstrated that adhering to usability guidelines increased efficiency which results in reducing organizational costs [50]. Such calculations and justifications are highly reminiscence of Taylor's justification for the application of scientific management.

It was only towards the end of the twentieth century that emotion has joined the classical consideration of physical and cognitive factors in HCI research and practice. The reasons for this delay are numerous (e.g., [4, 71, 73]), and are at least partially due to the nascent stage of the technology itself. The rapid progress of interactive technology, which has contributed to the development of more intricate and interactive products, has also contributed to the materialization of at least parts of the vision of ubiquitous computing. Thus, it is quite interesting to note that ubicomp has advanced simultaneously with the inclusion of emotions and aesthetic considerations in IT.

The recent increase in the HCI community's interest with the visual aesthetics of interactive systems overlaps the developments discussed above. It would be a mistake, though, to assume that only now have people began associating computing technology with aesthetics. For example, already in the early 1960s, computer scientists had experimented with the computer's ability to generate visual art [47]. It is quite clear that as hardware and software technology improved, so did the aesthetic opportunities afforded by it to designers and to users. However, it was not until very recently that visual aesthetics has become a mainstream topic in development and research of interactive systems. In the next chapter we discuss the aesthetic of technological products from three perspectives and describe how they relate to ubiquitous displays.

3 Perspectives on the Aesthetics of Interactive Products

In this section, we present three perspectives on the aesthetics of interactive technology, which we believe support our contention about the centrality of aesthetics to any form of interactive technology, and especially to such a salient technology as large situated displays. Thus, we look at aesthetics of interactive technology from a design perspective, from a psychological perspective, and from a practical perspective.

3.1 The Design Perspective

The activity of developing interactive technology is one of this era's most salient design activities. It is therefore important to evaluate this activity in terms of broader expectations from design activities and outcomes. In recent years, there has been considerable increase in both scholastic and business interest and recognition in activities, principles, and products, generally known as design (e.g. [11, 17, 37, 41]). This recognition should come as no surprise, as design is at the core of human activity, shaping every aspect of our lives. However, similar to the slow acceptance of the relevance of emotions and aesthetics in HF and HCI, the recognition that developing people-centered technology has much in common with other design disciplines has emerged only recently. It is not our purpose here to delve into the core questions that surround interaction design. Rather, we would like to point out two implications of handling interactive technology from a design perspective. First, is the recognition that aesthetics constitutes an integral part – in fact a cornerstone – of any design discipline. Second, aesthetics is often positively related to other design aspects – such as usability – which are at the core of the long standing tradition of HF and HCI.

3.2 The Vitruvian Design Principles

Information technology (IT) has much in common with aesthetics. IT scholars consider architecture as a reference discipline [5, 28, 33]. It is probably not a coincidence that the term "information architecture" has been adopted by professionals engaged in the activity of designing information-based environments and systems. As two of the most complex and intricate design disciplines, architecture and interactive technology share several important similarities. To exemplify the commonalities of the two disciplines, let us consider the Vitruvian principles of architecture. More than twenty centuries ago, Vitruvius argued that there are three core principles of good architectural work. *Firmitas*, the strength and durability of the building; *utilitas*, the utility of the building and its suitability for the needs of its intended users; and *venustas*, the building's beauty. In architecture, the Vitruvian principles have been influential since their rediscovery in the fifteenth century [26, 34]. For example, the Construction Industry Council in the U.K. has developed the Design Quality Indicator to evaluate the design quality of buildings based, to a large extent on Vitruvius's three main principles [16]. We believe that these principles make as much sense when designing interactive information technology.

Researchers and practitioners in the information technology fields quickly recognize the *firmitas* principle as one that stands at the core of their activities. Much of the traditional work in computing and informatics deals with this principle. It is impossible for a computer-based system, by its very nature, to function if it is not designed in a robust, stable and durable manner.

The second principle, *utilitas*, best describes the traditional practice and research in the field of HCI (cf. [72]). For example, one of the most widely referenced models on people's acceptance of information technology – the technology acceptance model (TAM) – deals with ease of use and usefulness as the antecedents of IT acceptance [10]. Similarly, efficiency measures were the main concern of seminal HCI research [6] and its offspring. Here we deal with how information technology should be designed to meet individual and organizational needs and goals. We are concerned with the efficiency and the effectiveness of using the interactive systems.

Yet, to complete the adherence of interactive technology design to the Vitruvian principles, we need to also consider *venustas*. Thus, beauty and delight are not a matter of face-lifting and make up. Aesthetics is at the core of good design practices, as can be seen in any other design discipline. The importance of aesthetics has now been recognized by HCI and HF researchers [28, 36, 71] and the reason for this should be obvious when we consider that interactive technology is all around us.

3.3 *Aesthetics Relates to Other Design Principles*

It is often the case that usability experts warn us against putting too much emphasis on aesthetics (e.g., [49]). A common sentiment among usability professionals is that, in the best case, usability and aesthetics are two different, unrelated things. In the worst case, they just cannot coexist. Your system is either beautiful or it is usable, but it can't be both. In our view, while design principles (such as Vitruvius's, but also many others) may be conceptually orthogonal, in practice they may be positively correlated. Research suggests that this is the case, for example, for perceptions of aesthetics and usability [8, 30, 31, 64, 69], albeit not under all circumstances [2, 12]. Our contention here is that not only are perceptions of aesthetics and usability often correlated, but so are the more objective design guidelines and criteria for design goodness.

This contention is based on conceptual and perhaps ideological arguments [77], according to which the aesthetic and the usefulness of information cannot be separated. Similarly, traditional, usability-based interaction design guidelines encourage order in design, keeping elements aligned, grouping various elements that belong together, clearly separating them from other elements, etc. These principles of gestalt perception are often mentioned as a blueprint for usable design. But these very same principles and guidelines can also serve us towards aesthetic design as well. Consider, for example the screens which appeared as Fig. 2 in Parush et al. [54]. The two screens were presented to participants in the study as representing two different levels of some quality of the design. The top screen represented good design, the left screen represented bad design. Can the reader guess whether that design quality was usability? Or was it aesthetics? The first author's experience in posing this question to different audiences result in practically the same distribution of answers: About one-third of the people surveyed vote for "usability"; another one-third vote for "aesthetics"; the remaining one-third vote for "don't know" or "both"!

Why do people respond like they do? Probably because it does appear that the best screen is superior both in terms of usability and aesthetics. The point, we believe, lies in the fact that there are many facets to aesthetics. Some of these aspects coincide with usability. For example, Lavie and Tractinsky [31] conducted a study on how people perceived the aesthetics of web pages. They found that users identified two general subdimensions of aesthetics. One of them was termed "classical", because it communicated a sense of order and good proportions. This subdimension was highly correlated with usability. The other subdimension was labeled "expressive" – it represents the originality and the creativity of the design. Its correlation with usability perceptions was much lower. We believe that when usability experts warn against overemphasizing aesthetics they refer to issues that might occur when the design is overly "expressive". Regardless of whether expressive aesthetics hampers usability (in our opinion it doesn't necessarily), we think that we should keep in mind that aesthetic is a complex concept, and at least some aspects of it can be viewed as enhancing, not contradicting, usability.

3.4 Implications for the Design of Ubiquitous Displays

Designing ubiquitous displays cannot afford to ignore aesthetics. The quest for aesthetics is embedded in Weiser's original vision of ubiquitous computing and is rooted in the basic principles of the design disciplines. Adherence to good design practices behooves us to consider how such displays appear to us, how information is displayed on them, how they are integrated with the environment in which they are embedded, and how they contribute to the aesthetic message that this environment communicates.

Designers of situated displays are right to emphasize the quality of the displays' content as their prime design objective [46]. They are also aware of the usability issues that accompany the content presentation [39]. However, we submit that taking into account the aesthetic aspects of the displays not only will not detract from the quality of the information presentation; to the contrary, it is likely to improve it.

4 The Psychological Perspective

Some in the IT community may consider aesthetics as gratuitous. This certainly used to be the sentiment during the early days of the discipline (e.g., [40, 51, 65]). However, as interactive technology becomes so embedded in our daily lives, it becomes important to reevaluate this position. In this section we argue that, from a psychological perspective, aesthetic design is important because it influences both emotional and cognitive processes [32]. Such influences affect people's experiences and attitudes towards the technology and what it represents (e.g., [21, 22]). In this chapter we first discuss the emotional and motivational aspects: aesthetics pleases us and improve our well-being. We then discuss cognitive processes by which visual stimuli are easily recognized and thus are essential to subsequent evaluation of products and environments.

4.1 Aesthetics Satisfies Basic Human Needs and Is a Source of Pleasure

While early (and to some extent current) thinking in the HCI community belittled the need to consider aesthetics, it is quite apparent that such a stance stands in contrast to what we know about human nature. In other words, we argue that aesthetics plays a significant role in motivating people and in satisfying their needs (e.g. [42, 58, 63]). Aesthetic design is a source of enriched positive experiences, pleasure and happiness [23, 24], although these experiences may vary, as people differ in their beauty evaluations [63] and are not equally sensitive to aesthetics [3], and obviously have different aesthetic tastes.

Given that we seldom run into perfect designs in terms of either usability or functionality, we can at least take comfort in the pleasures provided by aesthetics. Norman [53] argues that living in an aesthetic environment makes us more tolerant. Without aesthetic design we might be more impatient, less willing to forgive. Porat and her colleagues have demonstrated association between aesthetics and pleasure in online environments [55–57]. The influence of aesthetics on pleasure is important because happiness, in turn, has numerous positive effects on human well-being [38]. It is argued that pleasure changes the way we process information; it broadens the mind and improves creativity and divergent thinking [53]. We do not yet have conclusive experimental evidence that aesthetics improve performance, but some initial results suggest that it may. For example, in a recent study Moshagen et al. [44] found that aesthetic design enhanced performance under conditions of poor usability in the context of extracting health-related information from websites.

An interesting manifestation of how, when given a chance, people react to aesthetics of interactive technology can be seen in the ways that people modify their computer desktop and applications. Most new computing environments and applications come today with a personalization (or "skinnability") option, which allows users to change the appearances of those applications. For example, a music player will provide you with equal functionality even when it is dressed in various shapes and forms. You can download numerous skins from the web. Millions of such skins are being downloaded from major skin sites on the web, attesting to the growth of the demand for aesthetics of interactive applications [74]. Interestingly enough, studies have shown that the major factor influencing users' selection of skins was the aesthetic aspect of its design [70, 74].

4.2 Aesthetic Impressions Are Fast, Enduring and Consequential

A frequent argument for the importance of aesthetics is that our first impressions of objects or of other people are primarily visual and that consequently, these first impressions color the way that we perceive those objects or people. The phenomenon that people evaluate other people based on their looks was termed "what is beautiful is good" [13]. Research has documented numerous contexts in which people with good looks enjoy preferential treatment. While this is clearly an overgeneralization of the complexities of human-to-human interaction, and some of us may feel uneasy about its moral implications, there is some striking evidence to its existence (e.g. [19, 20, 60]). Similarly, we may also speak of a "beautiful is usable" stereotype, according to which beautiful products are considered by users to also be more usable (Tractinsky et al. 2000). Thus, at least under certain circumstances, more aesthetic systems were perceived to be more usable than less aesthetic systems.

The claim that aesthetic evaluations influence evaluations of other system (or person) attributes relies on the premise that aesthetic impressions can be formed faster than other forms of evaluations. This argument is based firstly on the quite obvious assumption that it is easier to observe aesthetic qualities of an object than more intrinsic qualities such as usability and functionality. Secondly, research has shown that aesthetic evaluations of objects can be made very quickly. Research suggests that people formulate impressions of visual beauty within around 300 ms [25]. In the context of evaluating web pages it was demonstrated that exposure of less than 500 ms had been enough to generate reliable and consistent aesthetic judgments [35, 71]. Thus, even before we interact with a system, our aesthetic perceptions of that system are formed and color our subsequent experience of using it.

4.3 Implications for the Design of Ubiquitous Displays

The appeal of beautiful environments, products or application has value to people [23, 43]. This appeal and its consequential value may depend on the objectives of the displays. It is probably more pronounced in entertainment, leisure, or other hedonic contexts than in contexts that are purely task oriented. Still, it is difficult to ignore the influence that large displays have on the environment in which they are embedded (e.g. Fig. 1). Studies have shown that aesthetic aspects of the environment influence people's attitudes towards them (e.g. [14, 18, 29, 48, 56]).

Digital displays also enable tailoring the messages of individual content providers and to different target groups. Additionally, live sensor data (like weather) and live Web data (like news or train delays) can be presented [45] facilitating the tailoring of information not only in terms of its content but also in terms of its aesthetic relevance to the requirements and expectations of varied audiences, depending on the timing and context of the presentation.

Emphasizing the need to design aesthetic displays is thus commensurate with the basic vision of ubiquitous computing. The design of information delivery objects and of the way information is presented by them will set the tone for people's experiences with ubiquitous displays. Aesthetics not only pleases the observer or the operator. It also affects perceptions of various attributes of the technology or of the organization that is represented by the technology [59]. Thus, perceptions of the content [22] presented by the displays, or the trustworthiness of the organization displaying the information [9] may also be affected by aesthetic aspects of the technology.

5 The Practical Perspective

We can also examine the importance of aesthetics IT from a more practical perspective. The goal of this section is to present arguments that go beyond the theoretical perspective presented in the two previous sections. Again, we do not

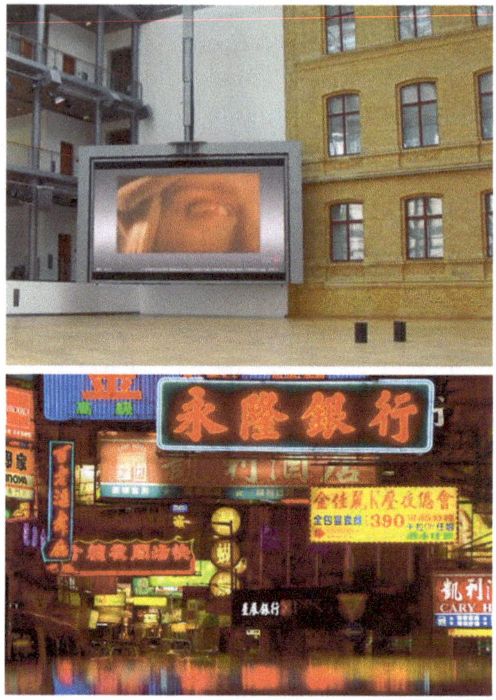

Fig. 1 Different atmospherics created by ubiquitous displays in Deutsche Telekom's Berlin Representative Office (*above*) and in Hong Kong

aspire to provide an exhaustive coverage of this perspective. Rather, the intention is to demonstrate some key arguments that stem from this perspective. We do so by arguing, firstly, for the importance of aesthetics as a differentiating factor among IT products. Secondly, we present the idea that integrating aesthetics into the design of information technology is an inevitable process that is already underway.

5.1 Aesthetics as a Differentiating Factor

As early as 1998, Norman [52] had outlined the transformation of much of the IT industry from that of technology-based products geared towards technology enthusiasts to one that is increasingly producing consumer products. Norman has argued about the business model of information appliances. Yet the essence of his arguments applies to various forms of interactive technologies, from cell phones to portable computers. We argue that as this process of consumer-centeredness and commoditization of interactive technologies becomes faster and more prevalent, the importance of aesthetics as a differentiating factor between competing products increases. A case in point is the digital watch industry, in which technology and performance have already met high enough standards of accuracy and reliability. In that industry, much of the differentiation between brands and models is now based on visual design. The situation is similar in many IT-based products. They are reliable; they provide much more functionality than what the average user needs, their prices drop consistently, and their usability improves. Thus, the competition becomes more oriented towards enhancing the user experience. Considerations of appearance and symbolic ownership are becoming more important than concerns for functionality, reliability, or even usability.

5.2 *Aesthetics Is Pervasive*

Even if we ignore all the aforementioned justifications for considering aesthetics in IT, it is impossible to ignore its pervasiveness both in terms of the aesthetic design of IT products and in terms of how IT produces aesthetics. Virginia Postrel [58] notes that one of the unintended and often overlooked characteristics of information technology is that it is particularly friendly to aesthetic applications. So users can relatively easily create, edit, transmit and receive aesthetic materials. A salient example for these qualities of information technologies can be found in the swift shift of the photography market from being completely analog (film) to being almost completely digital within about a decade. Digital photography is one of the great enablers of the phenomenon described by Postrel. The combination of digital photography with powerful application software has enabled designers in industries such as architecture, fashion, mass media, or Web development many more design options, and much more time to explore all of those options in order to create more appealing products. But, perhaps even more importantly, digital photography also offers ordinary people the chance to easily engage in creating and communicating aesthetics on a scale that would have been considered unthinkable in the past. Furthermore, as Postrel [58] convincingly argue, this trend has created a self-serving cycle: Being under constant supply of visually aesthetic stimuli, more people have become more sensitive to aesthetics; now they demand it more and expect it everywhere, including in ubiquitous displays.

Lest our position be misconstrued as placing practical concerns above all other considerations, we suggest that there is also a moral dimension in advocating aesthetic design. Aesthetic design implies that the designer or the organization respect their audience, were sensitive to their needs and desires and put effort into the design of the product and the environment (see Fig. 2). In turn, people will be more inclined to take care of an aesthetic and well maintained environment [62]. In short, aesthetic design works for the betterment of our lives.

5.3 *Implications for the Design of Ubiquitous Displays*

Designers of ubicomp environments and ubiquitous displays cannot remain outside contemporary approaches to the significant of aesthetic design. The value of

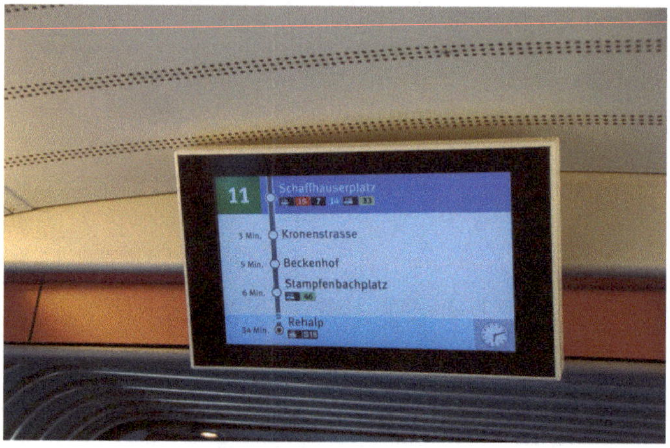

Fig. 2 Aesthetic design of a tram route display in Zurich, showing respect for the audience and the environment

aesthetic design is commercial, environmental, and personal. It touches on almost all aspects of our lives. Ignoring the need to carefully consider aesthetics may result in loss of business opportunities and in social rejection by a population that is increasingly conscious about aesthetic matters. Introducing aesthetic design where it didn't formerly exist (e.g. [61]) may create new opportunities. Designers of ubiquitous displays need to consider the aesthetic implications of their artifacts on the environment within which they will be embedded. They should also consider how the environment may affect perceptions of their products. For example, are the ambient conditions supportive or detrimental to certain modes of information presentation? In short, a plethora of consideration, some of which may be new to IT professionals, has to be seriously considered for projects involving ubiquitous displays.

6 Conclusions

In this chapter we have presented three perspectives from which the need for aesthetics in interactive technology is justified. The design perspective, the psychological perspective and the practical perspective each provide arguments for the aesthetic cause. Together, we believe that they form a solid justification for the growing awareness in many corners of the IT community about the importance of aesthetic design. Indeed, recent practice and research indicate that aesthetics become an integral aspect of interactive systems development. This process is by no means complete, and we hope that this chapter will contribute to clarifying its origins and support its continuance.

The implications of aesthetic design relate both to the physical aspects of ubiquitous displays, to their content, and to their integration with the physical and informational environments [7]. It is also important to note that while the aesthetic factor is important in influencing people's motivation, attitudes, and perhaps even behavior, these effects are not uniform. People differ in their aesthetic sensitivity and in their tastes. Their response to aesthetic stimuli may depend on the task, the type of information presented and probably other contingencies [66].

Many of the arguments put forward in this chapter are based on empirical findings from scientific and usability studies. We advocate that future research on ubiquitous displays will be similarly grounded in studies that would help tease the interplay between environments, displays and people. For this purpose there is a need to consider various aspects of potential interactions. For example, we need to identify relevant design characteristics of information displays; conceive typologies of use contexts, and measure potential effects and consequences of public displays installments and usage.

We have also discussed how aesthetic considerations apply to ubiquitous displays. We believe that there is strong connection between aesthetic principles that promote visual noise and clutter reduction (e.g. [75]) and the quest for calming technologies – one of the main motives for Wieser's vision of ubiquitous computing. Moreover, there is also a moral aspect to caring for the aesthetics of public spaces [62] that also coincides with ubicomp's vision that information technologies respect our private and public spaces. However, there seems to be a tension between the visions and principles of adequate, perhaps civic design on the one hand and practices of people who are in charge of developing our technological environment. It may therefore be prudent to embark on programs that improve visual literacy among IT students and practitioners. Tomorrow's environments are to large extent in their hands.

References

1. Bardram, J., and Friday A.: Ubiquitous Computing Systems. In: Ubiquitous Computing Fundamentals, pp. 37–94. CRC Press (2009)
2. Ben-Bassat, T., Meyer, J., and Tractinsky, N.: Economic and Subjective Measures of the Perceived Value of Aesthetics and Usability, ACM Transactions on Computer-Human Interactions (TOCHI), 13,2, 1–25 (2006)
3. Bloch, Peter H., Frederic F. Brunel, and Todd J. Arnold..: Individual Differences in the Centrality of Visual Product Aesthetics: Concept and Measurement. Journal of Consumer Research, 29, 551–565 (2003)
4. Brave, S. and Nass, C.: Emotion in Human-Computer Interaction. In: Jacko, J. & Sears, A. (eds.), Handbook of Human-Computer Interaction, pp. 251–271. Lawrence Erlbaum Associates, Hillsdale NJ (2003)
5. Brooks, F.P.: The Mythical Man-Month. Addison-Wesley, Reading, MA (1975)
6. Card, S.K., Moran, T.P., and Newell, A. The Psychology of Human-Computer Interaction. Lawrence Erlbaum Associates, Hillsdale NJ (1983)
7. Carter, S., Mankoff, J., Klemmer, S. R., and Matthews, T.: Exiting the Cleanroom: On Ecological Validity and Ubiquitous Computing, Human-Computer Interaction, 23, 1, 47–99 (2008)
8. Cawthon, N. and Vande Moere, A.: The Effect of Aesthetic on the Usability of Data Visualization. In: 11th International Conference Information Visualization (IV'07) (2007)

9. Cyr, D., Kindra, G. and Dash, S.: Website Design, Trust, Satisfaction, and E-loyalty: The Indian Experience. Online Information Review, vol. 32, 6, 773–790 (2008)
10. Davis, F.D.: Perceived Usefulness, Perceived Ease of Use and User Acceptance of Information Technology. MIS Quarterly 13, 319–339, (1989)
11. Design Council: The Impact of Design on Stock Market Performance: An Analysis of UK Quoted Companies 1994–2003, February (2004)
12. Diefenbach, S. and Hassenzahl, M.: The "Beauty Dilemma": Beauty is Valued but Discounted in Product Choice. In: Proceedings of the 27th international conference on Human factors in computing systems (CHI 2009), pp. 1419–1426 (2009)
13. Dion, K., Berscheid, E., and Walster, E.: What is Beautiful is Good. Journal of Personality and Social Psychology, 24, 3, 285–290 (1972)
14. Donovan, R.J., Rossiter, J.R., Marcoolyn, G. and Nesdale, A.:. Store Atmosphere and Purchasing Behavior, Journal of Retailing,70, 3, 283–294 (1994)
15. Fischwick, P. (ed.): Aesthetic Computing, MIT Press, Cambridge, MA (2006)
16. Gann, D.M., Salter, A.J., and Whyte, J.K. Design Quality Indicator as a Tool for Thinking. Building Research & Information, 31, 5, 318–333 (2003)
17. Gibney, Jr., F. and Luscombe, B.: The Redesign of America. Time (March 20, 2000).
18. Gilboa S., and Rafaeli, A.. "Store environment, emotions and approach behavior: applying environmental aesthetics to retailing", The International Review of Retail, Distribution and Consumer Research,13,2, 195–211 (2003)
19. Hamermesh, D., and Biddle, J.: Beauty and the Labor Market. American Economic Review, 84, 1174–94 (1994)
20. Hamermesh, D., and Parker, A.M.: Beauty in the Classroom: Professors' Pulchritude and Putative Pedagogical Productivity. Economics of Education Review, 24, 4, 369–376 (2005)
21. Hartmann, J., Sutcliffe, A., De Angeli, A.: Towards a Theory of User Judgment of Aesthetics and User Interface Quality. ACM Transactions on Computer-Human Interaction (TOCHI) 15,4 (2008)
22. Hartmann, J., Sutcliffe, A., De Angeli, A.: Investigating Attractiveness in Web User Interfaces. Proceedings of CHI 2007, April 28–May 3, pp. 387–396. San Jose, California, USA, (2007)
23. Hassenzahl, M.: Aesthetics in Interactive Products: Correlates and Consequences of Beauty. In: Schifferstain, H.N.J. and Hekkert, P. (eds.) Product Experience, Elsevier, Amsterdam (2008)
24. Hassenzahl, M. and Tractinsky, N.: User Experience – A Research Agenda, Behaviour & Information Technology, 25, 2, 91–97 (2006)
25. Höfel, L. and Jacobsen, T. Electrophysiological Indices of Processing Symmetry and Aesthetics. Journal of Psychophysiology, 21, 1, 9–21 (2007)
26. Johnson, P.A.: The Theory of Architecture: Concepts, Themes, and Practices. John Wiley & Sons, New York (1994)
27. Jordan, P.W.: Designing Pleasurable Products. Taylor and Francis, London (2002)
28. Kim, J., Lee, J., Han, K., and Lee, M.: Businesses as Buildings: Metrics for the Architectural Quality of Internet businesses. Information Systems Research, 13, 2 239–254 (2002)
29. Kluger, A. N., and Rafaeli, A.: Affective Reactions to Physical Appearance. In: Ashkanasy, N., Hartel C.E.J., and Zerbe W.J. (eds.).: Emotions and Organizational Life. Greenwood Publishing Group, Westport, CT (2000)
30. Kurosu, M., and Kashimura, K.: Apparent Usability vs. Inherent Usability: Experimental Analysis on the Determinants of the Apparent Usability. In: Conference Companion on Human Factors in Computing Systems, pp. 292–293, Denver, USA, (1995) http://portal.acm.org/citation.cfm?id=223355.
31. Lavie, T. and Tractinsky, N.: Assessing Dimensions of Perceived Visual Aesthetics of Web Sites, International Journal of Human-Computer Studies, 60, 3, 269–298 (2004)
32. Leder, H., Belke, B., Oeberst, A., and Augustin, D.: A Model of Aesthetic Appreciation and Aesthetic Judgments. British Journal of Psychology, 95, 489–508 (2004)

33. Lee, A.S.: Architecture as a Reference Discipline for MIS. In: Nissen H. E., Klein H. K, and Hirschheim R. (eds.), Information Research: Contemporary Approaches and Emergent Tradition, pp. 573–592. North-Holland, Amsterdam (1991)
34. Kruft, H.W.: A History of Architectural Theory: From Vitruvius to the Present. Zwemmer and Princeton Architectural Press, New York (1994)
35. Lindgaard, G., Fernandes, G. J., Dudek, C., and Brownet, J.: Attention Web Designers: You Have 50 Milliseconds to Make a Good First Impression! Behaviour and Information Technology, 25,2, 115–126 (2006)
36. Liu, I.: Engineering Aesthetics and Aesthetic Ergonomics: Theoretical Foundations and a Dual-Process Research Methodology. Ergonomics 46, 13–14, 1273–1292 (2003)
37. Lojacono, G. and Zaccai, G.: The Evolution of the Design-Inspired Enterprise. Sloan Management Review, Spring 2004, 75–79 (2004)
38. Lyubomirsky, S., King, L., and Diener, E.: The Benefits of Frequent Positive Affect: Does Happiness Lead to Success? Psychological Bulletin, 131, 6, 803–855, (2005)
39. Mankoff, J., Dey, A.K., Hsieh, G., Kientz, J., Lederer, S., and Ames, M. Heuristic Evaluation of Ambient Displays. Proceedings of CHI 2003, April 5–10, pp. 169–176. Ft. Lauderdale, Florida, USA (2003)
40. Marcus, A.: Graphic Design for Electronic Documents and User Interfaces. ACM Press, New York (1992)
41. Martin, R. and Riel, J.: Designing Interactions at Work: Applying Design to Discussions, Meetings, and Relationships. Interactions, March+April 2010, 16–19, (2010)
42. Maslow, A.H.: Motivation and Personality: Harper & Row, New York (1954)
43. Mehrabian, A. and Russell J. A.: An Approach to Environmental Psychology.: MIT Press: Cambridge, MA (1974).
44. Moshagen, M., Musch,, J. and Goritz, A.S.: A Blessing, Not a Curse: Experimental Evidence for Beneficial Effects of Visual Aesthetics on Performance. Ergonomics 52, No. 10, October 2009, 1311–1320 (2009)
45. Müller, J.H.: Context Adaptive Digital Signage in Transitional Spaces. Unpublished Ph.D. Dissertation, Westfälischen Wilhelms-Universität Münster. (2008)
46. Müller, J., Krüger, A. and Kufik, T.: Maximizing the Utility of Situated Public Displays. Adjunct Proceedings of User Modeling. Corfu (2007)
47. Nake, F.: Computer Art. A Personal Recollection. Proceedings of the 5th Conference on Creativity & Cognition. April 12–15, 2005, London, United Kingdom, pp. 54–62 (2005)
48. Nasar, J.L. (ed.): Environmental Aesthetics: Theory, Research, and Applications, University Press, Cambridge (1988)
49. Nielsen, J.: Usability Engineering. Academic Press, Boston (1993)
50. Nielsen, J.: Usability ROI Declining, But Still Strong. Available online: http://www.useit.com/alertbox/roi.html, Last retrieved August 15, 2010. (2008)
51. Norman, D.A.,: The Psychology of Everyday Things. MIT, London (1988)
52. Norman, D.A.: The Invisible Computer: Why Good Products Can Fail, the Personal Computer Is So Complex, and Information Appliances Are the Solution. MIT Press, Cambridge, MA (1998)
53. Norman, D.A.: Emotional Design: Why We Love (or Hate) Everyday Things. Basic Books, New York (2004)
54. Parush, A., Nadir, R., and Shtub, A.: Evaluating the Layout of Graphical User Interface Screens: Validation of a Numerical, Computerized Model. International Journal of Human Computer Interaction, 10, 4, 343–360 (1998)
55. Porat, T. and Tractinsky, N.: Affect as a Mediator between Web-Store Design and Consumers' Attitudes towards the Store. In: Peter, C. and Beale, R. (eds.) Lecture Notes in Computer Science: Affect and Emotion in Human-Computer Interaction, 142–153. Springer-Verlag, Berlin/Heidelberg (2008)
56. Porat, T. and Tractinsky, N.: It's a Pleasure Buying Here: The Effects of Web-Store Design on Consumers' Emotions and Attitudes. Human-Computer Interaction (in press).

57. Porat, T., Liss, R., and Tractinsky, N.: E-stores Design: The Influence of E-store Design and Product Type on Consumers' Emotions and Attitudes. Proceedings of the 12th International Conference, HCI International 2007, Beijing, China, July 22–27, 2007, pp. 712–721 (2007)
58. Postrel,V.: The Substance of Style. HarperCollins, New York (2002)
59. Rafaeli, A. and Vilnai-Yavetz, I.: Emotion as a Connection of Physical Artifacts and Organizations, Organization Science, 15, 6, 671–686 (2004)
60. Ravina, E.: Love & Loans: The Effect of Beauty and Personal Characteristics in Credit Markets. Columbia University (2008)
61. Reif, I., Alt, F., Hincapié Ramos, J.D., Poteriaykina. K., and Wagner, J.: Cleanly – Trashducation Urban System. Proceedings of CHI 2010, April 10–15, 2010, Atlanta, Georgia, USA, pp. 3511–3516 (2010)
62. Saito, Y.: Everyday Aesthetics. Oxford University Press, Oxford (2008)
63. Santayana, G.: The Sense of Beauty. Available online through Project Gutenberg: http://www.gutenberg.org/etext/26842 (Last accessed August 15, 2010) (1896)
64. Sonderegger, A. and Sauer, J..: The Influence of Design Aesthetics in Usability Testing: Effects on User Performance and Perceived Usability, Applied Ergonomics, 41, 403–410 (2010)
65. Spool, J.M., Scanlon, T., Schroeder, W., Snyder, C., DeAngelo, T.: Web Site Usability: A Designer's Guide, Morgan Kaufman, Los Altos, CA (1999)
66. Sutcliffe, A.: Designing for User Engagement: Aesthetic and Attractive User Interfaces. Morgan & Claypool Publishers Synthesis Lectures on Human-Centered Informatics (2010)
67. Taylor, F.W.: The Principles of Scientific Management. Chapters 1 and 2 available online: http://www.ibiblio.org/eldritch/fwt/ti.html (Last retrieved August 15, 2010) (1911)
68. Tractinsky, N., (1997). Aesthetics and Apparent Usability: Empirically Assessing Cultural and Methodological Issues. ACM CHI Conference Proceedings on Human Factors in Computing Systems (CHI 97), pp. 115–122.
69. Tractinsky, N., Shoval-Katz A. and Ikar, D.: What is Beautiful is Usable. Interacting with Computers, 13(2):127–145 (2000).
70. Tractinsky, N. and Lavie, T.: Aesthetic and usability considerations in users' choice of personal media players. Proceedings Volume 2 of the 16th British HCI Conference, London, September, pp. 70–73 (2002)
71. Tractinsky, N.: Towards the Study of Aesthetics in Information Technology, Proceedings of the 25th Annual International Conference on Information Systems (ICIS), Washington, DC, December 12–15, pp. 771–780 (2004)
72. Tractinsky, N.: Aesthetics in Information Technology: Motivation and Future Research Directions. In: Zhang, P. and Galletta, D. (eds.) Human-Computer Interaction in Management Information Systems: Foundations, pp. 330–347. M. E. Sharpe, Inc (2006)
73. Tractinsky, N., Cokhavi, A., Kirschenbaum, M. and Sharfi, T.: Evaluating the Consistency of Immediate Aesthetic Perceptions of Web Pages, International Journal of Human-Computer Studies, 64, 11, 1071–1083 (2006)
74. Tractinsky, N. and Zmiri, D.: Exploring Attributes of Skins as Potential Antecedents of Emotion in HCI in Fishwick, P. (ed.) Aesthetic Computing. MIT Press, Cambridge, MA (2006)
75. Tufte, E. R.: The Visual Display of Quantitative Information. Graphics Press, Cheshire, Connecticut (1983)
76. Weiser, M. and Seely Brown, J.: Designing Calm Technology. Xerox PARC. Online: http://www.ubiq.com/hypertext/weiser/calmtech/calmtech.htm (Last retrieved August 15, 2010) (1995)
77. Weiser, M. Gold, R., and Seely Brown, J.: The Origins of Ubiquitous Computing Research at PARC in the late 1980s. IBM Systems Journal, 38, 4, 693–696 (1999)

The Design, Deployment and Evaluation of Situated Display-Based Systems to Support Coordination and Community

Keith Cheverst, Faisal Taher, Matthew Fisher, Daniel Fitton, and Nick Taylor

Abstract In recent years, there has been much interest in the potential for situated displays to support sense of community. In this chapter, we describe our experiences of two significant situated display-based deployments which explore some of the issues that arise when such systems are used on a day to day basis. The first deployment described is that of the Hermes system which has been deployed and evaluated in both a University department and a domestic home setting. The second deployment is the Wray photo display which has been deployed in two locations in a rural village located in the north of England. Both deployments made strong use of user-centered design and have benefitted from longitudinal evaluation in order to provide key insights into issues of adoption and appropriation.

1 Introduction

In Weiser's seminal paper [23] introducing ubiquitous computing he described how:

> The most profound technologies are those that disappear. They weave themselves into the fabric of everyday life until they are indistinguishable from it.

This is perhaps the ultimate design goal for any piece of technology and the most challenging to achieve. So, in order for technologies to effectively disappear they need to be carefully designed such that they cause minimal disruption with the

K. Cheverst (✉) • F. Taher • M. Fisher
InfoLab21, Lancaster University, Lancaster, UK
e-mail: kc@comp.lancs.ac.uk

D. Fitton,
School of Computing, Engineering and Physical Science, University of Central Lancashire, Preston, UK

N. Taylor
Culture Lab, School of Computing Science, Newcastle University, Newcastle, UK

A. Krüger and T. Kuflik (eds.), *Ubiquitous Display Environments*,
Cognitive Technologies, DOI 10.1007/978-3-642-27663-7_7,
© Springer-Verlag Berlin Heidelberg 2012

existing practices, and so forth, of a given setting (such as a particular office environment/workplace or a family home etc.).

By depicting settings in this way, it is important to regard them as places and according to Harrison and Dourish [10] a place is:

> a space which is invested with understandings of behavioural appropriateness, cultural expectations, and so forth

Consequently, if the practices afforded or imposed by a ubicomp technology deployment do not fit in with associations of a given place which relate to cultural expectations, existing patterns of behaviour, etc., it is more likely that adoption of the technology will be problematic. As a designer/developer of ubicomp technologies it is therefore essential to understand the social and physical richness of a given setting. Typical approaches for understanding settings include ethnographic studies, use of cultural and technology probes, focus groups and design workshops. Iterative methodologies based on a cycle of observation, design and deployment are effective for understanding a given setting prior to deployment, for understanding the adoption of a deployed technology and the need for associated redesign/modifications.

Another factor which may increase the likelihood of successful adoption in a ubicomp deployment is the use of a design methodology that places a strong emphasis on end-user involvement – so called, user-centered design methodologies. Indeed, through approaches such as participatory design [1], the end users themselves may be strongly involved in the design process itself.

In the remainder of this chapter, we will focus on our research activities on exploring a particular kind of ubicomp technology – that of situated displays. When we refer to situated displays we agree strongly with the definition provided by O'Hara and his colleagues [15] and the possibilities they raise:

> At their most basic, digital display technologies allow information to be more easily updated dynamically and remotely. However, these new kinds of interaction technologies also allow people to use these situated displays in novel ways both as for the individual's purposes and in the support of group work.

While these displays are becoming increasingly common as the technology to realise them becomes trivial, most situated display deployments are simply passive advertisements showing slideshows of content despite the potential to offer novel interactions and real benefits to users. It is also important to note the focus on situated displays in this chapter which implies a need to take particular notice of the place of deployment when considering this particular form of digital technology.

In exploring situated displays we have three broad research aims:

1. *Understanding of settings.* In our research we use ethnographic and related studies (both longitudinal and short term) to understand the social nature of public and semi-public spaces both before and after the introduction of situated display technology. This work involved developing an understanding of the affordances of a given place (e.g. outside an office door or inside a communal living area) to help determine appropriate placement strategies for situated

displays and an appreciation of what content may be relevant to display in a given place to facilitate cooperation as well as a sense of community (defined as 'the perception of similarity to others . . . and the feeling that one is part of a larger dependable and stable structure' [18]) between and within a certain user group.
2. *Exploration of interaction and use.* Situated displays do not typically fit the traditional single user mouse/keyboard interaction style. Consequently, we sought to explore the interactions that manifest themselves (over time) in the domain. Much of this exploration was guided by our understanding of the settings and utilised techniques found in context-aware computing (location-aware behaviour, automatic personalisation/content creation based on sensed context, etc.) and tangible interfaces as well as more familiar modalities such as e-mail, instant messaging and mobile phones.
3. *Prolonged deployments.* A key element of our research methodology is the use of substantial deployed installations. The long term use of novel technologies, especially their collaborative and community effects, cannot be deeply understood through short-term experiments or 'toy' installations. This development and deployment enables longitudinal studies as well as being a technology demonstrator for dissemination and inspiration. It is also important to note that all the deployments described in this chapter run 24 h a day, 7 days a week.

Settings that we have studied as part of our exploration of situated displays have included: the homes of lecturing/research staff, a Computing Department space at Lancaster University, a residential care facility, a University climbing club and a rural village near Lancaster.

In the next section we present relevant related work, and then, in the remainder of this chapter, we focus on the situated display deployments associated with the Computing Department setting (Sect. 3), the homes of lecturing/research staff (Sect. 4) and a rural village (Sect. 5). Following these three sections, Sect. 6 provides a general discussion on the key challenges that we have encountered during our work on the design and evaluation of public displays. Finally, Sect. 7 contains our concluding remarks.

2 Related Work

In this related work section we summarise significant past and present research into the use of situated displays designed to facilitate notions of community and categorise this research according to four broad areas according to the type of deployment setting. Section 2.1 discusses deployments in research lab settings, Sect. 2.2 focuses on deployments in university/school settings, Sect. 2.3 focuses on deployments in so-called 'Third Places' [16] while Sect. 2.4 focuses on deployments in rural, urban or developing world contexts. Effectively, the ordering of these sections can be seen as representing deployments in increasingly 'wild' settings.

2.1 Situated Displays in Research Lab Settings

Some of the earliest work to investigate the potential of situated displays to support notions of community has taken place in research lab settings. For example, McCarthy's Groupcast system [14] utilised a large public display coupled with presence-sensing technologies in order to display content of mutual interest to work colleagues as they passed by the display. McCarthy also developed the 'OutCast' system which enables the owner of an office to display content (personal web pages, public calendar entries, etc.) on a medium-sized display touch-screen display situated outside his or her office.

Another piece of work involving a similar deployment in a research lab setting was the Plasma Poster Network [8]. This system sought to promote 'community information sharing' though a network of plasma displays which effectively acted as 'digital' poster boards. The system was deployed and evaluated over a longitudinal period of time (10 months) at the FXPAL research lab in California, U.S. and this enabled a number of ecologically valid observations to be made. This included observations regarding the appropriateness of content submitted for display.

Brignull and Rogers [2] have also studied public interaction around situated displays. One of the study settings used was a welcome party for postgraduates entering university in which projection based situated display system was used. The focus of their research was to:

> provide designers with a framework by which to understand and conceptualise public display-based interactions

For example, the research involved the use of video cameras to observe the movements of students around the display and so support the analysis of how students interacted with the display in this particular setting and event.

2.2 Situated Displays in University/School Settings

The Notification Collage (NC) [9] groupware system was developed, deployed and evaluated in a university setting by a research group at the University of Calgary. The system enabled distributed and co-located colleagues to post media elements, e.g. sticky notes or video elements, onto a real-time collaborative surface in the form of a large display in a public setting. The surface could also be viewed on personal private displays. Greenberg and Rounding comment that:

> we saw that people treated the NC as a virtual room/bulletin board encouraging interaction. One person would post a media element, and others would (eventually) become aware of it and selectively react to it.

In the school environment, the Dynamo multi-user situated display system developed by Izadi et al. [11] deployed and evaluated in a number of different settings including a sixth form common room. In this setting, it enabled students to submit media to the display and 'carve' specific areas of the public display for more private interactions.

2.3 Situated Displays in 'Third Places'

The term 'Third Places' refers to places which are not typically associated with home or work activities such as cafes and places where 'one may go alone at almost any time of the day or evening with assurance that acquaintances will be there' [16].

One important deployment in the café space was 'Cow Cam' [20] which featured a large plasma display accompanied by a web cam device and a flat bed scanner placed on one of the café tables. The 'Urban Grind' café which is the setting for the deployment is a coffee house located in Portland, Oregon, U.S. One significant finding of this work was the importance of supporting various methods of allowing patrons of the café to contribute content to the display. For example, patrons could use the web cam to take a picture of themselves or scan in an image of a leaflet they wished to advertise using the flatbed scanner.

McCarthy et al. [13] developed the Community Collage (CoCollage) system which was designed 'to cultivate community in a café', and features a display showing content such as photos and quotes uploaded by café patrons and staff. The authors state that the system provides:

> a new channel for awareness, interactions and relationships among people there.

Regarding the system's impact upon 'sense of community', McCarthy et al. observed:

> The CoCollage display offers a semi-public window into a community, revealing some of the interests – and interestingness – of members of the community in a place that serves as its hub.

Another piece of research in the cafe setting was carried out by Churchill et al. [7]. This research involved the design, deployment and evaluation of the eyeCanvas community display in a combined cafe and art gallery in San Francisco. One interesting aspect of this system was that the situated display was a touch screen and the eyeCanvas application enabled patrons to scribble and share messages in the form of 'finger scribbles'.

Kray et al. [12] also carried out a situated display deployment in the café domain (in the city of Newcastle in northeast England) which received longitudinal evaluation and revealed sense of community to be supported.

2.4 Situated Displays in Urban Settings

The CityWall system [17] comprises a multi-touch interactive public display deployed in Helsinki city centre. The large display (measuring 2.5 m wide) shows user generated photos of the city that have been tagged with 'helsinki'. Passers-by who chose to interact with the display can perform a variety of gestures to resize, rotate and reposition the photos on the display. The large size of the display means that multiple users can interact with the display and its content at any one time.

3 Hermes Displays: Supporting Coordination and Community in a Computing Department Setting

3.1 Overview

We have explored the use of situated displays within the Computing Department at Lancaster University in two generations of the Hermes door and photo displays. The first deployment of Hermes (Hermes 1) consisted of ten door displays and a photo display. Owners of the ten displays were split between lecturers, Ph.D. students, research assistants and administrative staff.

In June 2004, the Computing department moved to a new specially built building called InfoLab21. With this relocation we saw the opportunity to build on the results of the Hermes 1 installation with a new, larger, deployment of Hermes displays (Hermes 2). A total of 40 displays were deployed based on new hardware with larger screens. We also included a new photo display with this deployment.

3.2 Hermes 1 Door Display and Photo Display Deployment

This first generation was based on PDAs situated beside office doors. They allowed office occupants to set and display messages and visitors to scribble messages using an attached stylus. This deployment was active between April 2002 and July 2004, a total of 27 months. During this time frame over 5,278 owner messages were set and 775 visitor messages were left. Through these we gained a substantial insight into long-term acceptance and usage of semi-public displays along with context sharing behaviours and preferences in this environment. An example of these Hermes door displays installed outside offices in the computing department as shown in Fig. 1.

One of the questions we wished to answer with the first deployment of Hermes displays was that of whether traditional methods for sharing personal information and context, e.g. sticking a post-it note on your office door, could be achieved with a digital equivalent. Further to this aim, we wished to explore the new possibilities this digital equivalent would allow to extend these traditional methods into new realms, such as remote interaction and rich media messages.

3.2.1 Supported Functionality

In deciding the functionality to support in the Hermes display systems, a user centred approach was taken whereby functionality was added to the system through consultation with, and exploration of requests made by, the users of the system. Through this, our intention was to develop systems with simple and intuitive functionality easily accessible by a range of users (within the Computing

Fig. 1 An early Hermes 1 display

Department at Lancaster) with sufficient dependability to enable adoption and use over a longitudinal deployment.

The functionality supported by the Hermes 1 system can be considered from two distinct perspectives, namely: the perspective of the owner of the Hermes 1 display and the perspective of a visitor to the Hermes 1 display. Visitors were able to leave the owner a message by drawing with a stylus on the Hermes 1 display. Based on requests from owners we designed the system such that messages left by visitors did not remain on the screen but disappeared from the display once entered. An owner could read his or her messages left by visitors on a secure web portal. Owners were also able to configure textual messages and image files to appear on their Hermes 1 display using the web-portal. A typical textual message is shown in Fig. 2 and illustrates how an owner would often use their display as a means of sharing personal context based on either their current activity (as in this example), location (e.g. if the message had read 'Gone to Gym') or time (e.g. if the message had read 'Back in 5 minutes'). A full analysis of the ways in which the Hermes system was adopted and used by owners in order to share personal context information in order to support 'awareness' amongst colleagues is presented in Ref. [4]

Consultation with owners suggested the need for a feature to enable the owner to create a freehand message by using a touch screen interface on the door display itself. An example of the kind of message that was left on door displays once this feature was implemented is shown below in Fig. 3.

This feature required the addition of basic owner authentication, which was implemented with a simplified username/PIN system with an on-screen keypad. However, owners found the overhead of authentication too high and so following consultation we introduced a means for enabling a 'temporary' message, such as 'Gone for lunch' by tapping twice on their Hermes display. The first tap brings up a set of buttons each one representing one of a set of predefined messages. A second tap is then required to select their chosen message. This technique for setting

Fig. 2 A typical textual owner message left on a Hermes 1 display

Fig. 3 A typical 'scribbled' owner message

temporary messages proved very popular for the majority of Hermes owners. Indeed analysis of usage logs revealed that the average number of messages set per day tripled following the introduction of this interaction feature. This highlights an important trade-off between flexibility/control and effort on behalf of the user.

The system also enabled the owner to use SMS or MMS via their mobile phone in order to remotely send a message or picture to their Hermes display. This feature of Hermes proved extremely useful for enabling a door display owner to post their current status, for example, if stuck in bad traffic on the way to work or otherwise delayed from being at their office (see [5] for more details regarding this aspect of the system and its usage).

One of the most popular features to be added to Hermes (during the later stages of deployment) was one which enabled owners to e-mail messages to their displays. One owner in particular (a member of the senior administrative staff) had a regular

Fig. 4 The first Hermes Photo Display

habit of e-mailing a department-wide mailing list whenever she was going to be away from the office for more than a few hours. On discussing her requirements, it became evident that adding an e-mail feature to Hermes, allowing these messages to be CC'd to the Hermes system, would greatly reduce the effort required by her to keep her presence information updated on her door display. After implementing this feature the secretary made extensive use of it.

One common type of image that door display owners (in particular lecturers and researchers) posted to their door display was photographs of their latest conference venue. This observation led us to consider whether a larger situated display could prove useful that would support the display of such photos. In particular, we were interested in investigating whether or not by placing such a display in a corridor we would see community usage develop around the display by people with office situated close to the display.

Consequently, we developed and deployed the first Hermes Photo Display in a corridor on the lower floor of the Computing Department building (see Fig. 4 below).

The display was in place for a period a several weeks and at the end of this period we spoke informally to those people with offices on the corridor shared with the display. The feedback we received confirmed that the people on the corridor (mostly Ph.D. students but also some lecturers) all felt that the sense of community had strengthened in the corridor given the patterns of use that had developed around the display – namely, people on the corridor sending pictures to the display of places they had visited or humorous content.

3.3 Hermes 2 Door Display and Photo Display Deployment

Following the dismantling of the original Hermes 1 system in July 2004 and our move to a new department building in Infolab21, we realised an opportunity to

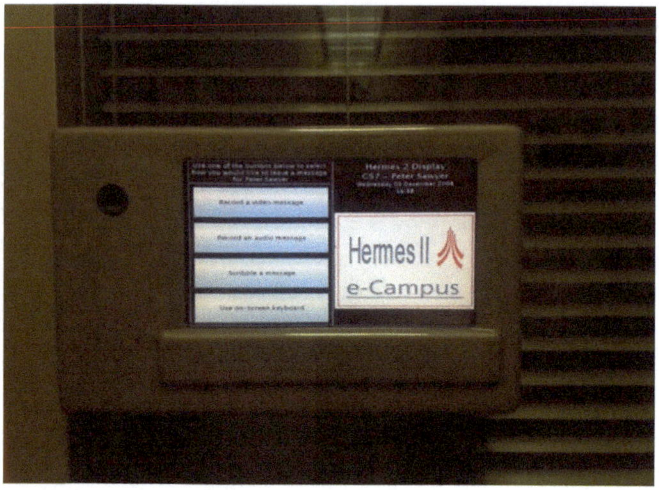

Fig. 5 A Hermes 2 display showing both owner message and visitor buttons

create an even more extensive ubiquitous display deployment. This consists of a full deployment of Hermes displays across 2 corridors and 40 offices. These new Hermes 2 units were designed based on multiple user studies and much consultation. They include cameras, microphones and Bluetooth, as well as the use of a larger 7 in. widescreen display. This larger screen was chosen by the majority of door display owners from the original Hermes system during a 'show case' study in which a variety of display options (based on high fidelity prototypes) were presented to previous owners.

The larger screen area has meant that door displays have enough screen real estate to enable the screen to be divided into 'visitor' and 'owner' sections (see Fig. 5). In the 'visitor' section, the owner can decide which of a set of messaging options are available to the visitor. For the door display shown in Fig. 5, the owner has chosen to have the following four messaging options available to any visitors: 'Record a video message', 'Record an audio message', 'Scribble a message' and 'Use on-screen keyboard'.

Alternatively, owners can decide to have the entire screen area reserved for their own messages, as shown in Fig. 6 below.

Many of the offices in the new Computing Department are multiple occupancy and Hermes 2 door displays now provide support for shared offices. In this case, the GUI of the display is divided into a number or rows with one row per person. For example the display shown below in Fig. 7 is used to support a shared office with two occupants.

The Photo Display was also deployed in the new InfoLab21 setting. The display itself was mounted using a wooden structure as shown in Fig. 8.

The new setting for the display was in a communal area open to a much wider audience than the previous (students, lecturers, administrative staff and research

Fig. 6 A Hermes 2 display showing the owner message only, in this case the owner has chosen not to have his door display support the functionality enabling visitors to leave messages

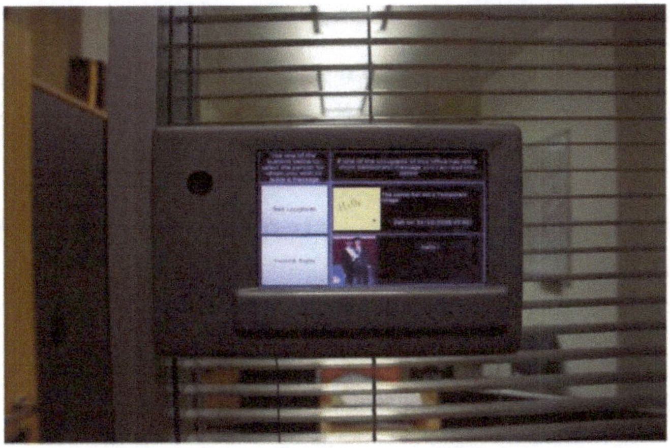

Fig. 7 A Hermes 2 display supporting a shared office with two occupants

assistants). Consequently, the application was modified to enable any user to submit or download content and to support multi-user interaction. In more detail, a user could use their Bluetooth enabled mobile phone to upload a picture to the display or to select and then download a photo image from the display to his or her mobile phone. The design and implementation of this system posed a number of technical constraints and interaction challenges (see [3] for a comprehensive description), e.g. how to manage the pairing of a specific user's phone with the photo display when multiple users are within Bluetooth range of the display. This particular challenge was solved by displaying (as a set of labelled buttons) the names of Bluetooth phones discovered by the Photo Display and prompting the user to select (by touching the appropriate button) their particular phone.

Fig. 8 A visitor to the Computing Department downloading an image from the Photo Display onto her mobile phone (March 2006)

An exploratory user study provided a number of useful insights into general user acceptability issues, interaction issues and the potential for such a display to facilitate notions of community. The user study involved a mix of 17 undergraduate and masters students and overall the response of students to the display was very positive. The results of this user study are described in detail in Ref. [3].

4 Hermes Displays: Supporting Intimacy in a Home Setting

The Hermes@Home system is a version of the Hermes system that has been tailored for deployment in the home. The layout of the display is similar to that of the Hermes 2 displays but instead of buttons being shown for leaving a message the 'scribble' message pane is always displayed. The basic idea behind the displays was to support notions of intimacy between family members when one of the members was away from home for extended periods of time. The Hermes@Home unit would act as a display for content, e.g. photos or text messages of the person away from home, while the at-home family members would be able to scribble messages on the units touch screen display in a very lightweight fashion.

A small number of initial 'formative' deployments have taken place and an analysis of use has revealed many similar categories of messages to those encountered with the Hermes deployments (see Saslis-Lagoudakis et al. [19] for more details). A typical home deployment is shown below in Fig. 9.

It is important to note that the placement of this Hermes@Home unit (as was the case with all deployments) was very carefully chosen by the family members. In more detail, its placement was at a high traffic location that was frequently passed by the various family members.

The main pattern of interaction that would occur around the display was one of the family member checking for new pictures or messages from her partner and

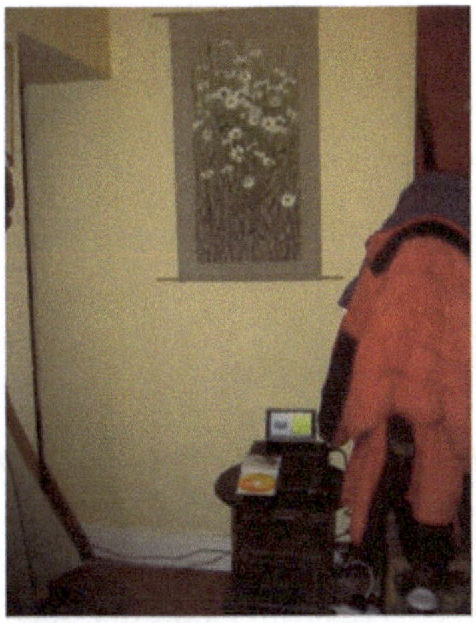

Fig. 9 A typical Hermes@Home deployment, situated in the home of a lecturer during a 6 week extended visit to Australia. The display is shown towards the *lower centre* of the photo

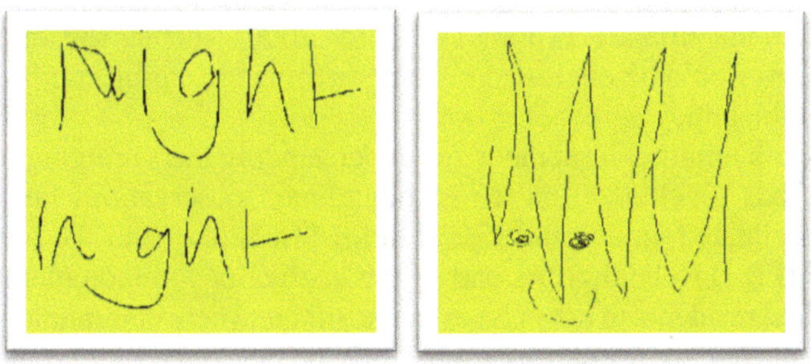

Fig. 10 Typical Hermes@Home messages left by the at-home family member

scribbling 'touches' of intimacy. Examples, of these kinds of messages are shown below in Fig. 10.

It is important to observe how expressive the scribble style of message leaving can be compared to, for example, a text message.

Interviews with the family members who had displays in their homes revealed how even though many of the messages scribbled onto the Hermes@Home display did not appear to contain much information or appear to require much effort to write, there were issues regarding the recipient (the 'away' family member) reading the messages in a timely manner. For example, one family member commented how they would not want to talk to their partner over the phone until they had had read all the messages that they had been left since the last time they had spoken. Another family member commented how: 'I think it would be good to see what [messages] the other person has read or not'.

The interviews with family members also revealed how the displays appeared to have embedded some form of presence of their partner. For example, one family member commented:

> For me it was a bit like a window to where [anon] is.

And another family member commented:

> When I left, I said goodbye to it as a link to you.

5 Situated Displays: Supporting a Sense of Community in a Rural Setting

Following our positive experiences with the Hermes Photo Display we decided to explore whether situated displays could facilitate and support a sense of community within a rural village setting. Lancaster's computing department had previously worked with members of a village called Wray (a small village not far from the university with a population of about 500) and so we decided to ask one of our contacts within the village (via e-mail in March 2006) whether she and some of her fellow villagers would be interested in being involved in a project utilizing situated displays, to which they agreed.

We utilised a number of methods for understanding the setting in Wray village. These methods included basic ethnographic observations of traditional noticeboards situated around the village (Fig. 11a below) and the use of cultural probe packs (Fig. 11b below). As part of the instructions for completing the probe packs we asked residents to note places in the village where community content was currently displayed and were it might usefully be displayed in the future.

We also attended community events to gain further insight into the domain and potential users such as the annual scarecrow festival in which members of the village put on public display scarecrows which they had designed and created themselves (Fig. 12).

Following the return of the probe packs, a meeting was held in the local village pub (Fig. 13a) in which it was decided to deploy a 'technology probe' in the village in the form of a simple digital display that would show pictures of the scarecrows from the annual scarecrow festival (the photos initially being uploaded by the author rather than a member of the village). This display would effectively be a stand-alone version of the Hermes Photo Display and it was agreed that it would be deployed during the summer of 2006 for several weeks in order to gauge initial community acceptance and feedback. Initially, we encountered some difficulties in communicating the concept and potential of the situated display and how the simple display of 'village pictures' could help foster a sense of community within the village. However, fortunately on display within the pub were a collection of historic photos of past village life (Fig. 13b) and, when these were highlighted to villagers at the meeting, the point regarding community was successfully made.

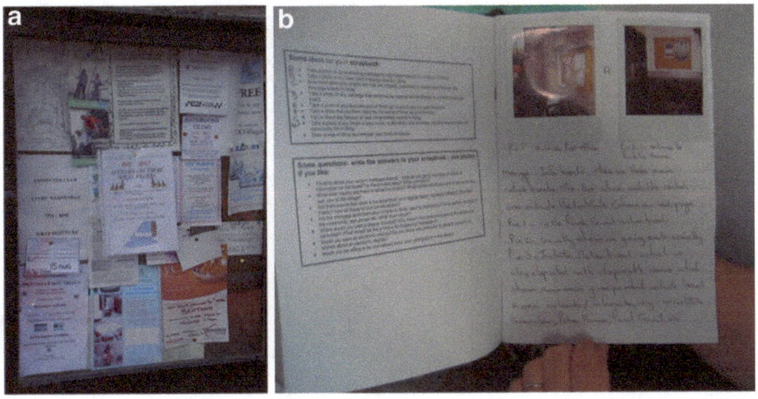

Fig. 11 (**a**) A noticeboard outside the Wray village hall, and (**b**) part of a completed probe pack

Fig. 12 Examples of scarecrows on display in the village of Wray during its annual scarecrow festival public event

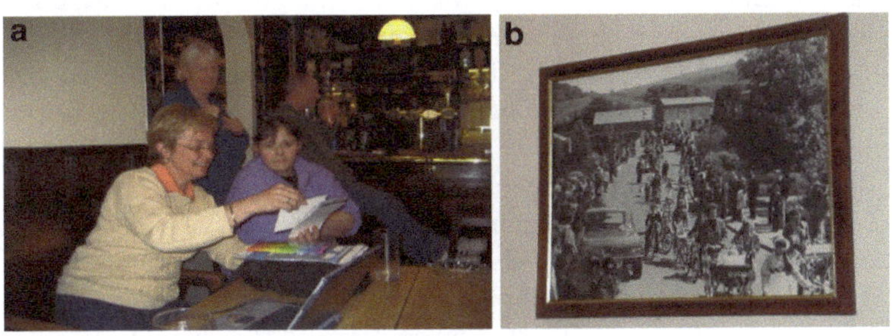

Fig. 13 (**a**) A meeting with residents in the village hall, and (**b**) historic photos of past village life on display in the village pub

Fig. 14 The technology probe being used by an elderly resident in the village hall

The photo display was also in place for a village 'produce show' event which took place towards the end of the summer and during this event the technology probe received significant usage and many comments were left by users. Figure 14 below shows the display being used by one of the village's elderly residents who was unable to identify herself in one of the pictures and is writing in the comments book a suggestion for the pictures to be made bigger to enable her to view them more clearly.

Following the produce show, the technology probe was refined using the requests for additional functionality (received either via e-mail or the comments book placed next to the display). The most significant change to functionality was support for additional categories of photos (i.e. in addition to the scarecrows category). However, supporting this additional functionality posed the question of whether or not to force some kind of moderation process in order to ensure the appropriateness of any and all new content that would be uploaded to the additional categories. Another meeting was held with villagers to discuss this issue and it was decided to support additional categories through the use of multiple moderators. In more detail, when a request was made for a new category of pictures, e.g. historic photos, then the villager making the request would be required to agree to moderate content for that particular category. This way it was envisaged that control of the system would be much more in the hands of the villagers as opposed to academics at the university.

The display remained in place for over 3 years, during which time various alterations have been made in response to feedback from residents and our own observations of the display in use. Many of these were minor alterations, including the ability to post comments on photos and to send digital 'postcards' by email, which were popular features suggested by community members themselves. However, the most substantial change came after feedback repeatedly suggested that residents would like to be able to post advertisements and news for local events and businesses.

After deciding to continue development of the display in this direction, a number of meetings were held to determine how this functionality would best be provided. Two events which had occurred in the village on the day of one meeting were particularly helpful in relating our discussion to real events: firstly, residents had

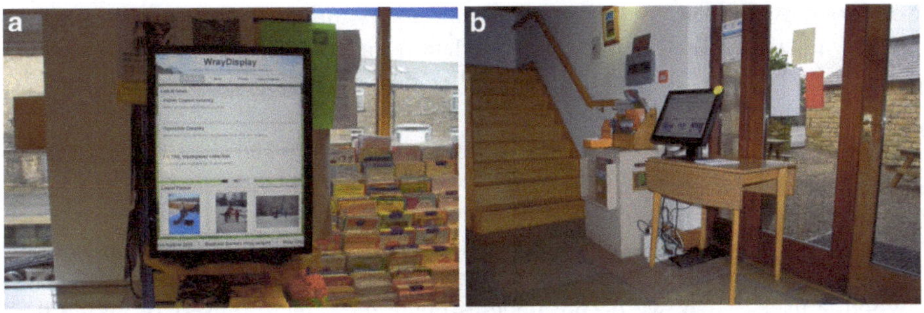

Fig. 15 (**a**) The WrayDisplay prototype in the post office, and (**b**) The WrayDisplay prototype in the cafe

missed a consultation about a proposed wind farm because the news had been posted a month earlier and forgotten, and secondly, a suspected scam artist had visited the village and persuaded an elderly resident to spend £1,000 on frozen fish. Discussion around these events helped to identify a need for breaking news and increased prominence for notices as they became more salient.

We also aimed to explore how much screen space for photos users were willing to sacrifice in exchange for news content. Although participants were extremely uncomfortable when asked to sketch user interfaces themselves, we subsequently found that pre-prepared paper mock-ups were useful for stimulating discussion around the user interface. Surprisingly, residents at the meeting supported a far greater emphasis on news than we had expected, stating that much of the photo content displayed was now several years old. Based on this feedback, WrayDisplay was deployed into the post office on a larger screen featuring both photographs and notices (Fig. 15a). The improved architecture of this new display also enabled the deployment of a second display into a small cafe/bistro in the village (Fig. 15b), with content shared across both displays.

The displays have now been operational in the village for a considerable period of time and further detail on usage and issues that have risen from the deployments can be found in Refs. [21] and [22].

6 Discussion

From the case studies we have observed that situated displays do indeed have significant potential to support notions of community. However, as discussed in Ref. [6]:

> in addition to the difficulty of successfully designing and implementing systems, the evaluation of their 'success', i.e. their ability to foster and support a sense of community, is also a challenging problem.

Our approach is to use a combination of quantitative and qualitative evaluation methods. The quantitative techniques analyzed usage logs while the qualitative techniques included: observation, informal feedback, interviews, questionnaires, focus groups, etc. Typically, we would use quantitative measures of use to inform qualitative evaluation, for example, to identify anomalies for further investigation and to validate qualitative findings. The data from these techniques has enabled us to continually collect and explore new/changing requirements in a timely manner during longitudinal deployment.

Additionally, evaluation has to 'fit in' with the users and the deployment (i.e. focus group/design workshop every week would not have been an option). The development team was then able to consider requirements on a range of levels (technical challenge, anticipated usage impact, novelty, alignment with aims of the work, etc.) and select features to explore through mote focussed design session involving users. Also, we do not underestimate the effect on users of taking their suggestions and feedback seriously and it has been important to communicate our commitment to them in this regard.

One crucial factor when designing situated displays to support community is that appropriate interaction methods should be in place to enable any user to create or alter content on their display (in other words, the system should consider the accessibility requirements of its entire user group). In the case of the Hermes Photo Display, situated in a communal area in the computing department of Lancaster University, we found that enabling users to upload content via his or her Bluetooth phone was a perfectly suitable method. However, despite some initial signs of enthusiasm from members of the Wray 'Computer Club', we found that Bluetooth was not a popular method for submitting content with the Wray Photo Display. Consequently, we introduced a website for allowing a greater cross section of the community to upload content and we are currently considering further approaches (e.g. the use of flatbed scanners) in order to further open up opportunities for posting content to the display by those members of the community who currently struggle with the website.

Another crucial factor for designers to consider is whether content moderation is required for the uploading of content and how this should be achieved in a scalable manner which ensures the community being served is in control. With the Hermes Photo Display the fact that users were required to register their phone's Bluetooth address acted as sufficient deterrent to prevent inappropriate content from being uploaded. However, with the Wray deployment it was necessary to implement explicit moderation processes in order to maintain the trust of the community.

7 Concluding Remarks

In this chapter, we have presented our exploration of the design, deployment and use of Situated Displays for supporting notions of collaboration (Hermes 1 and 2), intimacy (Hermes@Home) and community (Wray Photo Display). More specifically, we have described our work with three settings, the first that of a university campus, the second a home environment, and thirdly, a rural village in the north on England. Central to our approach is obtaining a strong understanding of the settings and the importance of actual deployments for understanding issues that can only arise through actual use in a real world setting and not through short-term experiments or 'toy' installations.

Our approach to evaluation is typically driven by the need to continually gather qualitative information from users both quickly and effectively in order to react to new and emerging requirements as swiftly as is feasible. As described throughout this chapter, we found this necessary in order encourage and support adoption during longitudinal prototype deployments. Crucially, in order to support coordination and community in our deployments it was necessary to weave feedback and participation from the community into a changing and evolving technological design.

References

1. Bjerknes, G., Ehn, P. and Kyng, M. (eds.) *Computers and Democracy - A Scandinavian Challenge*. Avebury, Aldershot, England. (1987)
2. Brignull, H. and Rogers, Y. Enticing People to Interact with Large Public Displays in Public Spaces, *in Proc. of INTERACT' 03* (Zurich, Switzerland), 17–24. (2003)
3. Cheverst, K. Dix, A. Fitton, D., Kray, C., Rouncefield, M., Sas, C., Saslis-Lagoudakis, G. and Sheridan, G. Exploring Bluetooth based Mobile Phone Interaction with the Hermes Photo Display, *in Proc. of the seventh ACM International Symposium on Human Computer Interaction with Mobile Devices and Services* (Salzburg, Austria), 47–54. (2005)
4. Cheverst, K., Dix, A., Fitton D., Graham, C. and Rouncefield, M. Situatedness of Awareness Information: impact on the design and usage of awareness systems, in: *Awareness Systems: Advances in theory, methodology and design. Springer HCI Series* – Eds: Panos Markopoulos and Boris de Ruyter, Chapter 17, 297–422, Springer-Verlag London Limited. (2009)
5. Cheverst, K., Dix, A., Fitton, D., Friday, A. and Rouncefield, M. Exploring the Utility of Remote Messaging and Situated Office Door Displays, *in Proc. of the fifth ACM International Symposium on Human Computer Interaction with Mobile Devices and Services* (Udine, Italy), LNCS 2795, Springer-Verlag, pp. 336–341. (2003)
6. Cheverst, K., Taylor, N., Rouncefield, M., Galani A. and Kray C. The Challenge of Evaluating Situated Display based Technology Interventions Designed to Foster 'Sense of Community', *in proc. of 2nd International Workshop on Ubiquitous Systems Evaluation (USE '08),* Graeme Stevenson, Steve Neely, and Christian Kray (eds.) (Seoul, South Korea, September 21, 2008), CEUR Workshop Proceedings, ISSN 1613–0073. (2008)
7. Churchill, E. F., Nelson, L., and Hsieh, G., Café life in the digital age: augmenting information flow in a café-work-entertainment space. In *CHI '06 Extended Abstracts on Human Factors in Computing Systems* (Montréal, Québec, Canada, April 22–27, 2006), CHI '06. ACM, New York, NY, 123–128. (2006)
8. Churchill, E. F., Nelson, L., Denoue, L., Murphy, P. and Helfman, J. The Plasma Poster Network: Social Hypermedia on Public Display, in *Public and Situated Displays: Social and Interactional Aspects of Shared Display Technologies* (eds. O'Hara, Perry, Churchill, and Russell), Kluwer Academic Publishers. Chapter 10, 233–260, July. (2003)
9. Greenberg, S. and Rounding, M., The notification collage: posting information to public and personal displays. In *Proceedings of the SIGCHI Conference on Human Factors in Computing Systems* (Seattle, Washington, United States), CHI'01. ACM, New York, NY, 514–521. (2001)
10. Harrison S and Dourish, P., Re-placing space: the roles of place and space in collaborative systems, in *Proc. of CSCW'96* (Boston, Massachusetts, United States), ACM Press, 67–76. (1996)

11. Izadi, S., Brignull, H., Rodden, T., Rogers, Y. and Underwood, M., Dynamo: A public interactive surface supporting the cooperative sharing & exchange of media. In *Proceedings of UIST'03*, (Vancouver, Canada). (2003)
12. Kray, C., Galani, A. and Cheverst, K., Engaging with Cultural Content on Ambient Displays, Urban Screens 2007, Manchester, UK. (2007)
13. McCarthy, J. F., Farnham, S. D., Patel, Y., Ahuja, S., Norman, D., Hazlewood, W. R., and Lind, J., Supporting community in third places with situated social software. In *Proceedings of the Fourth international Conference on Communities and Technologies* (University Park, PA, USA, June 25–27, 2009), ACM, New York, NY, 225–234. (2009)
14. McCarthy, J., Providing a Sense of Community with Ubiquitous Peripheral Displays, in *Public and Situated Displays: Social and Interactional aspects of shared display technologies.* O'Hara, K., Churchill, E., Perry, M., Russell, D. and Streitz, N.(eds). Chapter 12, 283–308, Kluwer. (2003)
15. O'Hara, K., Churchill, E., Perry, M., Russell, D. and N. Streitz, Public, community and situated displays: Design, use and interaction around shared information displays. *Workshop at CSCW 2002* (New Orleans, Louisiana, United States). (2002)
16. Oldenburg, R. and Brissett, D., The Third Place, *Qualitative Sociology*, Vol. 5, No. 4, 265–284. (1982)
17. Peltonen, P., Kurvinen, E., Salovaara, A., Jacucci, G., Ilmonen, T., Evans, J., Oulasvirta, A., and Saarikko, P., It's Mine, Don't Touch!: interactions at a large multi-touch display in a city centre. In *Proceeding of the Twenty-Sixth Annual SIGCHI Conference on Human Factors in Computing Systems* (Florence, Italy, April 05–10, 2008), CHI'08. ACM, New York, NY, 1285–1294. (2008)
18. Sarason, S. B. *The Psychological Sense of Community: Prospects for a Community Psychology.* Jossey-Bass. (1974)
19. Saslis-Lagoudakis, G., Cheverst, K., Dix, A., Fitton D., and Rouncefield, M. Hermes@Home: Supporting Awareness and Intimacy between Distant Family Members, *in Proc. of International Conference of the Australian Computer-Human Interaction Special Interest Group (OzCHI'06).* (2006)
20. Sherry, J., Beckwith, R., March, M., Salvador, A., and Barile, S., The life of the place: technology and communities. In *HCI International 2005* (Las Vegas, Nevada, United States). (2005)
21. Taylor, N., and Cheverst, K. Social interaction around a rural community photo display. *Int J. Hum.-Comput. Int. Vol. 67*, No. 12, 1037–1047. (2009)
22. Taylor, N., Cheverst, K., Fitton, D., Race, N.J.P., Rouncefield, M., and Graham, C. Probing communities: Study of a village photo display, in *Proceedings of OZCHI'07* (Adelaide, Australia), ACM, 17–24. (2007)
23. Weiser, M., The Computer for the 21st Century, Scientific American, Vol. 265 No. 3, 66–75. (1991)

xioScreen: Experiences Gained from Building a Series of Prototypes of Interactive Public Displays

K.P. Ludwig John and Thomas Rist

Abstract xioScreen stands for a series of prototype installations of interactive public displays that have been developed, deployed, and evaluated at Hochschule Augsburg over the last 5 years. On the one hand xioScreen installations are part of an emerging digital signage system that informs students and campus visitors about events, talks, etc. On the other hand, xioScreen installations serve as testbeds for exploring new forms of interactive entertainment. Our initial working hypothesis was that students would take advantage of public screens as shared playgrounds for new gaming experiences. In terms of user participation, however, this did not work out so far – therefore other forms of active user participation have been explored as well. In this paper, we report on a number of conducted design-deployment-evaluation cycles and discuss lessons learned so far.

1 Introduction

Increasing affordability of large plasma screens and more recently LCD screens attracts the signage industries as well as researchers interested in exploring new ways of informing and entertaining people in public spaces, be they indoor or outdoors.

K.P.L. John (✉)
Department of Graphics Design, Hochschule für angewandte Wissenschaften Augsburg, Augsburg, Germany
e-mail: john@hs-augsburg.de

T. Rist
Department of Computer Science, Hochschule für angewandte Wissenschaften Augsburg, Augsburg, Germany
e-mail: thomas.rist@hs-augsburg.de

A. Krüger and T. Kuflik (eds.), *Ubiquitous Display Environments*, Cognitive Technologies, DOI 10.1007/978-3-642-27663-7_8,
© Springer-Verlag Berlin Heidelberg 2012

Companies[1] promoting digital signage solutions usually point out that their technology is much more flexible than conventional signage, as their technology can cope with all kinds of media-rich digital content, and also facilitates editing and instantaneous distribution of new content.

Researchers – currently especially those working in computer science disciplines, such as HCI and pervasive computing – have recognized public screens as a stimulating element for a broad variety of research strands to be explored. Research is often much technology-oriented, e.g., the focus might be on the development of mechanism for intelligent screen layout, or automated content capturing from online sources. Another strand of research aims at investigating usage patterns and social implications of public screens that support some sort of active user involvement, e.g. by means of private access/control devices, such as smart phones or PDAs. Finally, a great deal of research and development activities aims to bring about new applications that foster group communication in the work place, as well as new applications for the entertainment of possibly anonymous users roaming around or dwelling in open public spaces.

In this paper we report on our so-called xioScreen system of which we have built a series of prototypes over the last 5 years at HSA (University of Applied Sciences Augsburg). From a research point of view our initial motivation was to enable interplay between shared public displays and private handheld devices, and to study how such technological settings would be used by students in daily life in a university campus environment. In addition, there was the more pragmatic consideration to furnish our university campus with a network of public information points that features large interactive screens for casting contextually relevant information.

2 Related Work

Our xioScreen prototypes share similarities with a number of systems that (a) deploy public displays for situated information display, (b) support some sort of shared user activity, such as gaming, or voting, and (c) enable user interaction via a handheld device, such as a smartphone. The terms "Shared Public Displays", "Interactive Public Screens", "Ambient Displays", or in outdoor settings "Urban Screens" have been coined and are often used in the literature to refer to systems of this kind [11, 13]. A comprehensive overview on the state-of-the-art in this area can be found in the Ph.D. thesis by Jörg Müller [11]. His survey reveals that the majority of systems has been developed within the last decade only, and that most of them are research prototypes which only faced very limited exposure to target user groups. For the purpose of this article, we restrict ourselves to a short number of references which we found of particular relevance to our own work.

---

[1] E.g. see product information by InfoScreen AG, Wall AG, NAVORI SA.

Location and target audience: So far, our series of xioScreen prototypes has been targeted towards indoor usage at a university campus, assuming that the majority of the target users will be people populating a university campus. In this respect, we share similar goals with the e-Campus system at Lancaster University [4, 16, 17], and the iDisplays system at the University of Münster [7, 10, 11].

Public screens as testbeds for new interactive experiences: While public screens are often designed as mere information/ information sharing systems, we are also interested in exploring the potential of public screens in the development of interactive entertainments. For instance, similar to work by other groups [5, 9, 14, 15, 19] our screen installations can serve as play grounds for shared game playing experiences, or they may take on a role as part of an interactive art installation [16].

Interaction via Bluetooth enabled mobile devices: From a technical point of view the prototypes that we have built so far rely on Bluetooth enabled mobile devices to support interactivity [1]. This includes up- and download of messages as well as using Bluetooth-enabled devices as game controllers for multi-player games.

Longitudinal field trials: Designing new interactive technologies for durable deployment and large audiences is a difficult matter. It is relatively easy to build a prototype with some new innovative features but, without longitudinal deployment and evaluation, it is hard to make any conclusive statements on system usage and acceptance. Like in Refs. [3, 4, 7, 12, 18] we are interested in a real durable deployment of our prototypes. We expose them to a large audience (currently HSA students), and conduct periodical evaluations and feedback gathering which help us guide our further developments.

3 xioScreen Version 1

In 2006 we made a first attempt towards the development of an interactive public screen system. The initial vision for this endeavor was based on the assumption that users would use such a system heavily as an interactive platform:

- For playing multi-user games that are specifically designed for interactive public screens which are scattered over the university campus, and
- As media for a lively information exchange

These assumptions seemed reasonable as the target user group for the xioScreen system installations on the HSA university campus is formed by students in their early twenties, most of them being mobile phone users, and a considerable number of them play computer games, at least occasionally.

3.1 Design of xioScreen V.1

In the design phase of xioScreen V.1, a group of 16 students from the computer science and graphics design faculty worked out a number of different game

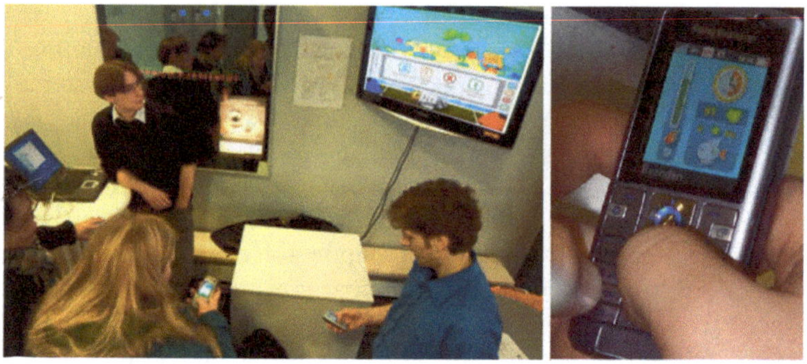

Fig. 1 Playing the aquarium game Makai. Fishes shown on the large display are controlled by users with mobile phones

concepts that they would like to play given a shared public screen to which users could connect using their BT-enabled mobile phones. Among the proposals they selected two of them which they found most engaging and suitable to be played in a public environment.

"*Makai* – friss Dich bunt" is a multi-user game that features an aquarium populated by fish characters which are "owned" by the players. Using a little Java ME application on a BT-enabled mobile phone, users new to Makai can join in by first creating their own fish and steering it in the aquarium which is shown on the public display. Of course, big fish eat small fish, so the challenge for the user is to escape from being hurt or eaten by bigger fish, and in turn, try to catch and eat smaller ones. The longer a user's fish manages to survive and eat other fish, the more points a user can collect. A table of user scores is accessible on a user's mobile phone, and a top ten list is occasionally shown on the public screen, too. The simple logic of game play allows users to compete with others, or likewise, users can form teams, e.g. depending on faculty membership. The screenshots in Fig. 1 give an impression of playing Makai. The two active players in the foreground of Fig. 1 use mobile phones to create characters (fish) which will appear on the public screen.

Makai has been designed as a durable, never ending game. Once a fish has been released into the aquarium, it will be around until it gets eaten by an enemy. If a human owner disconnects her mobile phone from the public screen her fish goes in sleeping mode (i.e. it disappears from the public screen) but will awake (i.e. appear) as soon as her owner connects again. No big tragedy if a user's fish gets eaten, only all previously collected points are lost but the user could simply create a new fish and try her fortune again.

TugWar, in contrast, was designed as a multi-user game to be played quickly by two competing teams of users. TugWar has been inspired by the classical tug-of-war game which pits two teams against each other in a test of strength, i.e., each team towing at the opposite end of a rope. In our TugWar implementation the members of the two competing teams connect to the public screen and receive little quests on their mobile phones. The more right answers are given by a team, the better the overall performance of that team. The actual team performance is visualized on the public screen using the "towing rope" metaphor.

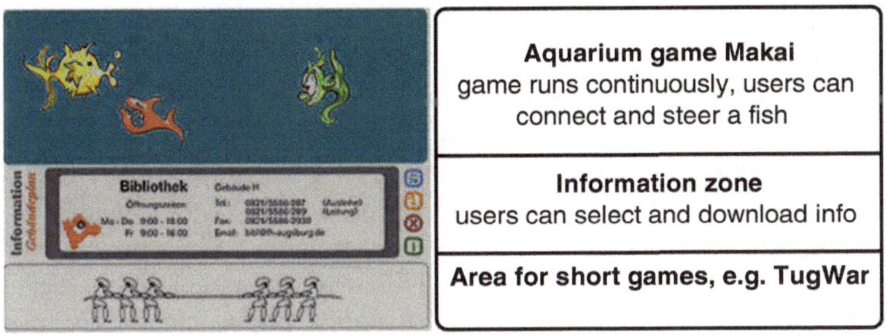

Fig. 2 Screen layout of the 1. prototype

Information zone: In addition to the two game applications, xioScreen V.1 comprises an area containing entry points for accessing campus-related information. By means of her mobile phone, a user can inspect and download news and messages which might be of relevance to her. Information items of xioScreen V.1 were restricted to campus information, such as upcoming events, campus maps, and opening hours of university facilities and offices.

Apart from designing user interfaces for each of the above mentioned applications it was a particular challenge to develop a layout that could accommodate all of them. We had to address issues, such as whether or not to enable simultaneous play of Makai and TugWar, and how to partition available screen real-estate among the applications. Eventually, we decided to use a split-screen layout for xioScreen V.1 reserving approximately half of the screen space for Makai, and each with a 1/4 for TugWar and the Information Zone (cf. Fig. 2). Also, trials on simultaneous playing of Makai and TugWar revealed quickly that a multi-game playing mode would rather confuse users. Therefore, games were selected by users on a first-come, first-serve rational.

3.2 Implementation Notes Prototype 1

The applications (Makai, TugWar, campus information service) comprised by our xioScreen V.1 prototype follow a classical divide between client-side and server-side modules. Client-side modules basically constitute the user interfaces to enable interaction with the public screens, and to access server-site information repositories. Mobile clients are implemented in Java ME for execution on Bluetooth-enabled mobile phones. Server-side modules are meant to provide the business-logic of an application. For instance, in the case of the Makai there is a server-side module responsible for updating the world state of the Makai aquarium according to steering commands received from a user. In addition a server-side application module may require a public screen as an output media, e.g., to display a campus map in response to a user request.

Since from the beginning it was assumed that additional applications will be developed, an attempt was made to factor out system parts which can be shared by several applications, and a plug-in mechanism was designed to facilitate the addition of further applications. In short, the overall system architecture consists of a basic messaging infrastructure, a so-called Plug-in Manager, and a set of server-side application modules. As applications, such as the Makai game, may rely on third-party components, such as the Macromedia's Flash player technology, they are wrapped by a piece of Java code. This way, an application is registered by the Plugin Manager and controllable via a mobile phone client. The messaging infrastructure keeps track of the connections with different mobile clients and handles message exchange between server-side application modules and a server-side Display Manager module that coordinates output from different sources on one or several public screen devices.

Users don't need to create accounts or reveal their identity when connecting via a mobile phone. However, MAC addresses of connected mobile devices are cached and used to distinguish between first-time users and returners, e.g. in the Makai game.

3.3 *Evaluation of Prototype 1*

The xioScreen V.1 prototype was ready for testing in early 2007. At an open-day event it was shown and demonstrated to a broad audience. Since a prerequisite for full system usage was that the client-software had to be installed on a user's mobile phone. At the open-day event, this was facilitated by human assistants who also explained users how to play the available games.

Before installing the system at the campus for permanent usage we asked a team of sociologists from the University of Augsburg to evaluate the prototype with regard to usability and potential acceptance. As a first step, they recruited a group of 50 students who filled out a questionnaire eliciting information on daily mobile phone usage habits, and expectations associated with an interactive public screen system. In a second step a small sub-group was invited to the laboratory to try out the prototype while being observed. This exercise was especially helpful for detecting usability issues of the mobile user interfaces.

Regarding expectations almost all subjects had an open and positive attitude towards the system and thought that they would use it. Those who interacted with the system appreciated the information services (opening hours of campus facilities, maps, etc.) but, to our surprise, were less enthusiastic regarding the provided game applications. Rather than a platform for gaming, the test users perceived the system foremost as an interesting option for receiving campus information, and when asked what would constitute desirable extensions, most of them made suggestions for additional information categories. Figure 3 summarizes the received feedback on querying users' preferences for public screen content.

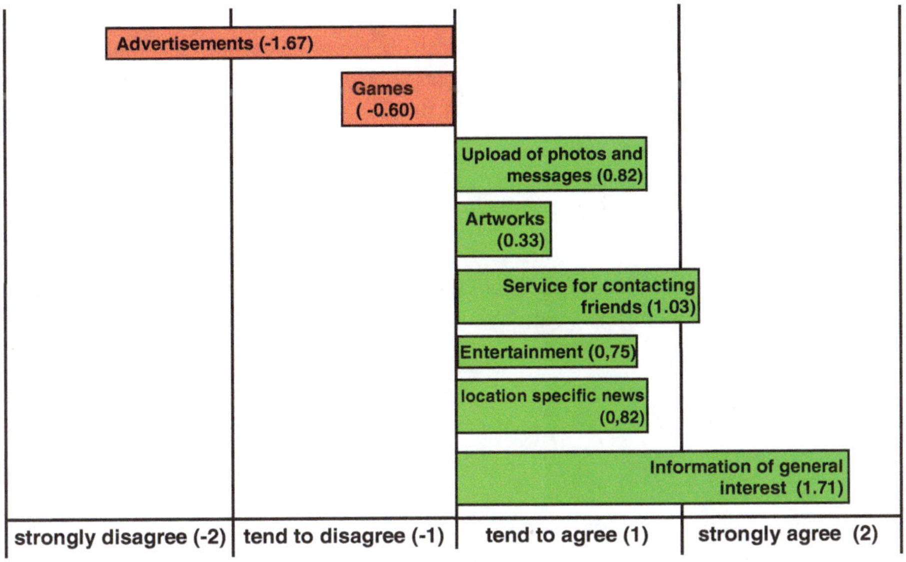

Fig. 3 Preferences for display contents on a public screen (Translated from Ref. [8])

However, the study was not sensitive to specific screen locations, and therefore did not reveal potential differences of user preferences at different locations.

3.4 Lessons Learned

After some polishing, a first screen installation was put into operation at a student's café on the HSA campus. The launch was accompanied by a number of promotional activities to draw the students attention to the interactive features of the system. After several weeks of operation a feedback questionnaire was distributed to students dwelling at the location of the screen installation. In addition, a built-in logging mechanism allowed retracing unattended interactive system usage. The lessons learned from returned questionnaires, log-file inspections, and also personal impressions gathered at the screen location can be summarized as follows:

The presence of a public screen system at the student's café was generally appreciated, especially if it would not display advertisements but contextualized (i.e. campus-specific) information. However, many students were not aware of the fact that they could interact with the system via a mobile phone. Not surprising, the log-files revealed very low interactive system usage.

There was also great reluctance of users to play Makai or TugWar. The few recorded gaming activities could be allotted to situations in which students who were involved in the xioScreen development brought friends to the screen location and invited them to participate in the game play. Of those users who actually tried to interact with the system, some reported that they suffered from occasional unstable

Bluetooth connectivity or failure to establish a connection at all. In sum, low participation in games was due to technical hurdles (i.e. the need to download software and problems with Bluetooth connectivity), but also – as revealed by the conducted survey – due to a generally low want for of playing games while at university.

4 xioScreen Version 2

Lessons learned from Prototype 1 guided the development of our Prototype 2. However, several design-test-evaluate cycles were carried out.

4.1 Prototype 1.5

In June 2007 a revised version of Prototype 1–Prototype 1.5 – was installed at the HSA Mensa. In contrast to our first system, more emphasis was put on information delivery. Based on a further user survey carried out among HSA students a number of information categories of high interest for students were identified:

- Information about HSA events
- Information on cancelled lectures
- Information about public transportation (e.g., delayed trains, trams, buses)
- Offers by students for used text books
- Offers by students for rooms and apartment sharing, and sharing trips by car or train.

A visible consequence of the new focus was a change in the screen layout. This time more space was devoted to the display of information categories. The area for "short games" (cf. Fig. 1) was dropped in favor of an enlarged information zone. Under the hood, a number of further improvements concerning bug fixes have been made, too.

This intermediate prototype was still based on the assumption that content, such as information about events, offers, etc. would be uploaded by by-passers via their mobile phones. It turned out quickly that this is not the case. One reason was certainly that students who were willing to upload information needed to have a BT-enabled mobile phone with them, and worse, if they were first-time users, they had to visit the xioScreen project home-page (www.xioscreen.de) first to download the client-software on their mobile phones. But even those students who made this step reported that they are still quite reluctant to act as content provider for various reasons (lack of time, lack of ideas what could be shown, lack of confidence that their messages would be of sufficient interest for others, and, last but not, aphaty).

4.2 Content Is Key

At this stage of the project, it had become quite clear that a more dramatic redesign of the overall system would be necessary in order to increase its attractiveness. If

Fig. 4 Messages are entered into the xioCMS and can be previewed by authors

Fig. 5 Shots of the dot-cube series to entertain screen onlookers (Cartoons created by Michael Maria Tichy)

for our target users not gaming but content is a key success factor, and if the students themselves would not "feed" the system, then we would need (a) to recruit reliable content providers/authors, and (b) as an enabling prerequisite, equip the system with an easy-to-use backend for authors.

Web Interface to an xioCMS: A major extension of the xioScreen concept was the addition of a web-based content management system (xioCMS). Registered authors of the xioCMS are able to edit messages, set display parameters (target screen location, first-time display date and expiry date, additional resources, e.g. a background image), and preview them before publishing. Figure 4 shows a screenshot of the web-based xioCMS authoring interface.

Recruiting content providers: In our search for suitable content providers, we contacted various bodies of the HSA administration as well as organizations which are affiliated in one or the other way with our university. Among those parties who were granted access to our web-based xioCMS the following became reliably active content providers:

- Student Services office provided daily updates on the menus served in the Mensa, as well as information on special campus events and activities;
- Catholic University Community (KHG) informed about their activities on campus and at their student dorm;
- Faculty secretariats mainly posted announcements of external guest speakers;

- Lecturers of several graphics design classes submitted diverse content produced by students as part of their class works. Examples include campaigns for events, continued cartoon series (e.g. the dot-cube series, cf. Fig. 5), short animation/stop-motion clips, etc.

4.3 Interactivity: Ballots Instead of Games

While feedback from students revealed that most of them enjoyed watching others playing Makai or TugWar, they showed great reluctance to play. In the case of TugWar, the barrier was even higher as playing required to form small teams, too. Eventually, it was decided to try out another form of user involvement, i.e. casting of votes. To this end, the xioCMS was extended so that authors could configure polls and ballots on a selected themes. For instance, the student services office suggested that students may indicate preferences on special meals that they likewise could put on the next weeks menu card. The university priest thought of having the students decide on the topic of his next sermon. Also, a built-in voting mechanism could be used for entertainment purposes, such as posing quiz-for-fun questions, or asking whether soccer team A or team B will win the upcoming match.

Ongoing ballots are announced on the public screen. Users willing to cast a vote can do so by using their BT-enabled mobile phones. As in the case of Makai and TugWar, a prerequisite for participation is, however, that a piece of software – the xio-mobile client – has been downloaded and installed on the mobile phone.

In order to draw the students' attention to the newly available voting feature, a cartoon-style animation clip was crafted and occasionally displayed on the screen as a kind of house advertisement.

As pointed out earlier, users do not have to create accounts or register. Since our voting mechanism just checks the MAC addresses of the connecting mobile devices, it could be easily fooled. To cast more than a single vote a user could simply use different mobile phones. However, this was not considered a severe problem as our polls/votings/quizzes are neither part of scientific studies, nor do they impact any critical decision making (like, e.g., elections of representatives would do).

4.4 Design Considerations for Prototype 2

Compared to the previous prototypes V.1 and V.1.5 we have completely changed the screen layout as well as the presentation strategy for displaying information items (i.e. news, announcements, etc.). Now approximately three fourths of the screen real-estate is used for the display of content, while the remaining one fourth is reserved for the announcement of ballots (cf. Fig. 6).

For the display of information items we adopted a rolling banner, or as we call it, a "carousel-style" approach. Information items have display items associated which are thought of sitting in an endless turning vertically carousel, while the screen

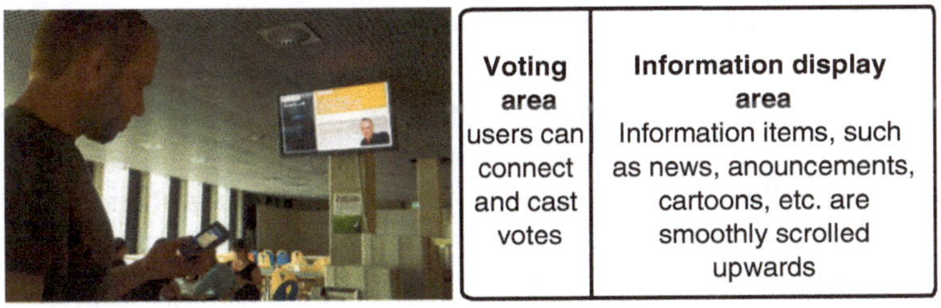

Fig. 6 Installation of a xioScreen V.2 at HSA Mensa and screen layout scheme

provides a view on two items at an instance in time. After a new item comes into sight at the bottom of the screen, the above display items are scrolled upwards until the new item is fully visible (and the top most item has disappeared), then the carousel stops a few seconds before it moves on (cf. Fig. 6). The restriction of showing only two items aside is to ensure sufficient readability of text messages, and also due to aesthetical considerations.

In case a display item takes on the form of an image (poster) or a video clip, we switch to full-screen mode and display it. However, after a maximum of 12 s the screen is switched back to carousel mode (i.e. as a consequence, video clips must not be longer than a maximum of 12 s). This value reflects a subjective threshold which has been determined in an informal experiment with a group of students who had to rank the appropriateness of different display durations shown on our screen at HSA Mensa. However, determining an ideal display duration for items, such as video clips is a difficult issue since it depends on the screen location, too.

Another tricky issue is the determination of a suitable display sequence for items in the carousel. There are several obvious rules for content scheduling that may apply, such as "avoid that an item is followed by itself", and "ensure that an item has a fair chance to get displayed at all". As suggested in Ref. [17] time filters may be used increase the priority of items with regard to a certain peak time, e.g., in timetables of trains and buses may be more relevant around the rush hours. Another rationale is to give priority to "urgent" items i.e. those which are getting close to expire date. We also discussed with our content providers (cf. Sect. 4.2) whether we should allow authors to assign priorities to information items at authoring time. Since there was the fear that potentially all content providers would consider their bits as of highest priority, we did not implement such a feature in the xioCMS user interface. In the discussion, an acceptance issue became also very clear that any automated scheduling mechanism – regardless of how sophisticated it may be – it must be transparent to our content authors how the mechanism works. Eventually, we ended up using a fairly simple mechanism to calculate an item's display priority. We consider an item's "urgency" (difference of expire date and actual date), a "display count" (how often has the item already been displayed), and "display pause" (how many other items have been shown since last display of the item under consideration).

4.5 Choosing Locations for xioScreens

The location as well as the positioning of our first public screen in the students' café wasn't optimal. The screen was placed on a wall closed to the entrance at a height of 150 cm above the floor. Students entering the café had to turn their heads aside in order to look at the screen. For the second prototype we did some in-situ pre-testing with paper mock-ups of a screen snapshot at potential locations and analyzed them with regard to criteria, such as:

Highly frequented dwelling zones: The conception of our second prototype assumes that users are willing to spend some time in front of its public screen. Firstly, information is displayed sequentially and some of the display items (video clips, animations) are timed media as well. Secondly, participation in a voting requires users to inform themselves about the options and to perform a sequence of steps for casting a vote. This suggests that locations should be preferred where many people go but also dwell for a while, e.g., a café area, a foyer, or a waiting area. In contrast, hall-ways and thoroughfares often do not satisfy this criterion.

Visibility: Ideally, a public screen is placed so that it is visible from a broad viewing angle and the screen will not be blocked by onlookers. In contrast to the first prototype, we placed the screen this time at a height of 1.90 m measured from the floor. Also, lighting conditions need to be considered, e.g., if placed close to a window sunlight may severely impact visibility of screen content.

Potential to catch peoples' attention: Ideally, a screen catches the attention of people entering the room/space in which it is installed. For instance, in a room with only one entrance door the screen has probably more potential to catch attention if placed at the wall opposite to the door. Also, the screen should not be placed too close to "competing" objects, e.g., paintings, pin boards etc. (see also Refs. [6, 20]).

Of course, there are many other factors that need to be considered, e.g. availability of power supply and Internet connection, safety aspects (e.g. the screen should not be placed too close aside emergency exit signs, and must not cause blocking of passage ways), and protection from theft and vandalism.

Our first choice for a screen location was the HSA Mensa area. There, we installed a large screen (actually a 46 in. television) so that:

- Students who line up in front of the food outlay will face the screen, and
- About a one third of students sitting in the eating zone have good sight to the screen too.

In addition to the Mensa area, we installed public screens at three further locations, all of them are basically foyer-style areas but feature a smaller number of visitors than the Mensa area.

4.6 Participatory Art Campaigns

Another form of user engagement was tried out by students of a graphics design class in 2009. In the spirit of participatory art, their aim was to involve visitors of

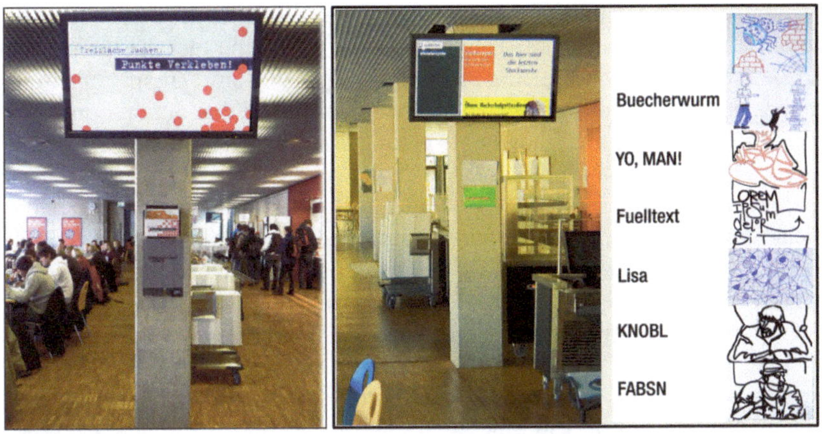

Fig. 7 Participatory art campaigns: on the *left*: "Punkte verkleben" (by Michael Cral), and "xioScrapper" (by Moritz Kartheuser) on the *right*. The drawings on the *very righthand side* represent stories of the emerging "skyscraper". The growing stack of stories was shown on the screen throughout the duration of the campaign

the HSA Mensa as co-creators in a creative process of making an artwork. The screenshots of Fig. 7 give an impression of the two conducted campaigns "Punkte verkleben" and "xioScrapper". In both cases the local public screen installation at HSA Mensa was an essential part of the artistic concept.

"Punkte verkleben" invited participants to express themselves by pasting red dots on paper posters. The role of the screen was primarily to animate Mensa visitors to participate. The campaign was run over a period of 3 weeks in January 2009.

For the "xioScrapper" campaign, the screen was also used to show the state of the emerging skyscraper based on contributions by students. Students could layout and furnish single stories of an overall skyscraper on picture postcards which were distributed at the screen location. Stapling these cards formed the "xioSkyscraper" of which a manually scanned version was shown on the screen. The xioScrapper campaign run over a period of 3 weeks during the summer semester in 2009.

4.7 Feedback on Prototype 2

Since spring 2008, prototype 2 is in daily operation during the semester turns but usually gets switched off during holiday periods. In 2011 we operated installations at four different locations (HSA Mensa, HSA portal, HSA foyer of auditorium, KHG dormins). Feedback on prototype 2 came from different sources:

Log-file data: Recorded activities of the xioCMS web-service reveal that on average between 10 and 30 information units per day are ready for display. Though content authors can activate location filters for their entries, they hardly make use of this feature (i.e., information items are shown at all four locations). Images and video clips are rarely provided. This is certainly due to the fact that preparation of such content is much more time consuming than writing a mere text message. Similarly, votings are rarely posted. One reason might be that our authors were less creative than hoped in finding suitable themes for votings.

Logged connections of mobile-phone clients reveal that when there was a voting issued, participation in it was very low, anyway. In sum, compared to the two game applications of prototype 1, the voting feature did not increase user participation.

Feedback from students: In 2009 a further questionnaire was issued to the students. Part of the questions was on what students liked about the current system. Another part of questions concerned the students' attitudes towards some suggested extensions. In addition to the questionnaires we received comments from students when discussing xioScreen issues as part of course-works in some lectures on HCI and interaction design. In sum, the great majority of students appreciated the new screen layout and the carousel-style information presentation.

In general, the redesign of the system was appreciated in terms of its modified visual appearance. Many people noted that the new design very much increased readability of posted information items. The adoption of a carousel-style information display strategy for xioScreen V.2, and the fact that more screen real estate has been reserved for displaying information items increased the perceived utility of the system. Though most of the students had a neutral or even positive attitude towards the voting feature but would hardly use themselves.

The two conducted participatory art campaigns "Punkte verkleben" and "xioScrapper" were quite successful in terms of achieved user participation. For example, in the case of the campaign "Punkte verkleben" a total of 1,750 red dots has been displayed for usage nearby the screen installation. By the end of the campaign, no dots were left but only 1,005 were pasted as intended on posters by Mensa visitors (the missing other dots were found elsewhere or got taken away as souvenirs). Taking into account the observed number of dots pasted per person (~10–20), and an estimated total of about 1,000 Mensa visitors during the duration of the campaign, we estimate that around 50–80 Mensa visitors (approx. 5–8% of all Mensa visitors during the campaign period) did actively participate in the creation process. Of course, this estimate includes some "returners" too, i.e. students who did paste dots on different days, however, we consider this as a success, too. Gathered feedback from Mensa visitors who did not paste dots themselves indicated that most of them found the campaign amusing and stimulating for their lunch talks.

In the case of xioScrapper we received an almost similar participation. At the end of the campaign we counted a total of 117 cards submissions with drawings of stories on it. Comparing the cards we estimated that around 70–80 stemmed from different authors.

Feedback from content authors: Feedback from active content authors was primarily targeted towards improving usability and functionality of the xioCMS. For instance, a message preview option was integrated based on suggestions from authors.

5 Towards xioScreen 3

While xioScreen 2 is operating smoothly and suffices its purpose as a visually appealing campus information system, there are many options to explore further, e.g. in order to increase active user participation. In a successor project which is conducted with Stadtwerke Augsburg[2] our focus lays on the following aspects:

Outdoor Screens for the General Public: xioScreen 2 is very much geared towards students populating the HSA campus. In contrasts, a screen at a tram connection station features a much more heterogeneous target audience comprising pupils, university students, business commuters, shoppers of all age groups, pensioners, tourists, etc. A first in-situ analysis of a selected tram station in the city center of Augsburg suggested that it would be beneficial to associate screen messages with a more fine-grained temporal filter, similar as suggested in Ref. [17].

It also deserves mentioning that going outdoor imposes a number of further considerations, such as protection from vandalism and impassiveness of weather conditions. Solutions for such problems exist but it is primarily a matter of available budget which may also impact decisions on choosing a suitable location.

Automated compilation of screen content from third-party sources: Our current xioCMS assumes that messages to be displayed on the screens are typed in by a human author. Though an experienced author may require less than a few minutes to log into to the CMS, enter a message, configure display parameters, preview the message, and release it for publication, it is often considered as an additional effort in their daily business. In some cases, it would be possible to skip manually entry of data since they are already available in digital form, e.g., the daily menu card of the Mensa is taken from an Excel sheet used by the student services office for planning anyway.

The straightforward idea is to program a wrapper that automatically accesses a data base or an online information source, collects and extracts information of interest, and feeds its outputs into the xioCMS. In our current project we explore two different strands:

Firstly, we are interested in how far the audience of public screens will appreciate messages (Tweets and Feeds) captured from Twitter and RSS streams. The interesting question is to find out which streams would be suitable for which kind of target user group.

Secondly, our project partner Stadtwerke Augsburg has a strong interest to inform citizens about various activities, e.g., production and supply of green energy, fresh water consumption/saving, number of trams and buses currently in operation, etc. To this end, we plan to compile animated information charts that will receive live data from the IT-center of Stadtwerke Augsburg. The ultimate communicative intent behind these charts is to improve the image of Stadtwerke as an

---

[2] Stadtwerke Augsburg is a public-owned incorporation responsible for the town's public transportation system as well as energy and fresh water supply, and sewerage.

Version	main Features	Layout of public screen
Prototype 1 (2006-2007)	• interactive games: Makai, TugWar • information zone, display on request only	Aquarium game Makai Information zone short game TugWar
Prototype 1.5 (2007-2008)	• interactive games: Makai only • extended information zone, display on request only	Aquarium game Makai Information zone
Prototype 2 (2008-2011)	• carousel-style display of information units (text, images, video clips) • interactive votings	Voting area Information display area
Prototype 3 (since March 2011) to be continued	• card-metaphor for the display of information units (text, images, video clips) • display of Tweets and Feeds	logo banner Information display area footer

Fig. 8 Evolution of the xioScreen prototypes from 2006 to 2011

innovative actor in its business areas. Technically, this requires an extension of our xioCMS so that an author can select a chart type and configure display parameters. As can be seen in Fig. 8, changes to the screen layout will be made as well.

Encourage users to interact: After several attempts to increase active user participation – e.g., users as gamers, users as authors, users as voters, users as down loaders of screen messages – we must admit that the number of interacting users is still disappointing. One reason for this is certainly the somewhat cumbersome need to download and install the BT-client software on the mobile device. In our next version we will replace the Bluetooth connection of mobile phones to public screens by a WLAN connection. This way we hope to lower the technical barrier for by-passers to become not only passive onlookers but active users, too. Further ideas to increase active user participation comprise the deployment of sensors, e.g., a camera that would enable a public screen to "see" whether a person is looking at it and use this information to splice in an instruction of how to use a mobile phone to interact with the xioScreen system.

6 Conclusions

Figure 8 summarizes the evolution of our xioScreen prototypes since 2006. In brief the experiences and insights gained so far from our installed systems at the HSA campus are as follows:

- In our initial vision an interactive public display would provide both, a campus-related information service as well as a playground for casual gaming. Up to now, this vision did not yet became a reality. Our screen installations are primarily recognized by their users as a visually appealing campus information system. However, to be appreciated as an useful system too, it must be fed daily with a sufficient amount of relevant and interesting content units, and display them in a way that attracts people's' attention (see also Refs. [6, 20]). To this end we had to recruit reliable human authors from different areas, and to provide them with an easy-to-use content management system, too.
- Motivating users to take advantage of interactive features appears to be more difficult than expected, and this applies for all our previous attempts that envisioned BT-enabled mobile phones as interaction devices (interactive games, ballots, up- and download of screen messages).
- In contrast, our two conducted participatory art campaigns received higher attention and were more successful in terms of achieved active user participation. One reason for this was certainly that participation in a creative design process was possible without bothering about any technologies (apart from pasting physical dots on paper, or drawing with a pencil on cards). Another reason might be that most Mensa visitors had no previous experiences with participatory art campaigns and therefore many of them were curious to find out more about it.
- As others suggested, too, e.g. Ref. [2], we think that the technical barrier should be as low as possible in order to transform passive onlookers into active doers. Assuming that more and more mobile phones provide easy-to-use access to the web, we will switch to WLAN access and web-based user interfaces for our next attempts on providing users with access to interactive public screen applications.

Acknowledgements The work on the xioScreen project has benefited from numerous contributions of several student projects run between 2006 and today as well as from colleagues who helped to supervise these projects. Special thanks to Andreas Romer and his colleagues for evaluation works on prototype 1, and to Marcel Kieser, Gerhard Meixner, Alexander Lawrence, and Stefan König for distinct technical contributions to the prototypes. We are also grateful to our external supporting project partners KHG, Studentenwerk, and Stadtwerke Augsburg. Also thanks to Elisabeth André for her comments on an earlier version of this report.

This article has been submitted for publication in Kuflik, T., and Krüger A. (eds.) "Advances in Ubiquitous Display Environments", Springer, forthcoming.

References

1. Ballagas, R., Borchers, J., Rohs, M., and Sheridan, JG.: The smart phone: A ubiquitous input device. In IEEE Pervasive Computing, 05(1):70–77, 2006.
2. Brignull, H., Rogers, Y.: Enticing people to interact with large public displays in public spaces. In Proceedings of INTERACT-03, pp. 17–24, 2003.

3. Churchill, E.F., Nelson, L., Denoue, L., Helfman, J., and Murphy, P.: Sharing Multimedia Content with Interactive Public Displays: A Case Study. In Proceedings of Designing Interactive Systems (DIS), 2004.
4. e-campus project website: http://ecampus.lancs.ac.uk/
5. Finke, M., Tang, A., Leung, R., and Blackstock, M.: Lessons learned: game design for large public displays. In Proceedings of the 3rd international Conference on Digital interactive Media in Entertainment and Arts (DIMEA '08), vol. 349. ACM, New York, 2008.
6. Huang, E., Koster, A., and Borchers, J.: Overcoming assumptions and uncovering practices: When does the public really look at public displays? In Proceedings of Pervasive 2008, 2008.
7. iDisplays project website: http://ifgi.uni-muenster.de/0/idisplays/
8. John K.P.L., Romer, A.: Vom Handy auf den Großbildschirm – Einsatzmöglichkeiten interaktiver Public Screens am Beispiel des xioSCREEN der HS. In Forschungsbericht 2009 der Hochschule Augsburg, pp. 91–96, 2009.
9. Leikas, J., Stromberg, H., Ikonen, V., Suomela, R., and Heinila, J.: Multi-user mobile applications and a public display: novel ways for social interaction. In Proceedings of Pervasive Computing and Communications (PerCom '06), pp. 66–70, 2006.
10. Müller, J., Paczkowski, O., and Krüger, A.: Situated public news and reminder displays. In Proceedings of European Conference on Ambient Intelligence, pages 248–265, 2007.
11. Müller, J.: Context Adaptive Digital Signage in Transitional Spaces. Ph.D. Thesis, Westfälische Wilhelms-Universität Münster, 2008.
12. Müller, J. and Krüger, A.: MobiDiC: Context Adaptive Digital Signage with Coupons. In Proceedings of the European Conference on Ambient intelligence. Lecture Notes in Computer Science, vol. 5859. Springer-Verlag, Berlin, Heidelberg, 2009.
13. O'Hara, K., Perry, M., and Churchill, E.F. (eds): Public and Situated Displays: Social and Interactional Aspects of Shared Display echnologies. Kluwer Academic Publishers, Norwell, MA, USA, 2004.
14. Reeves, S., Benford, S., O'Malley, C., Fraser, M.: Designing the spectator experience. In Proceedings of the SIGCHI conf. on Human factors in computing systems, ACM Press, pp. 741–750, 2005.
15. Scheible, J., Tuulos, V. H., and Ojala, T.: Story Mashup: design and evaluation of novel interactive storytelling game for mobile and web users. In Proceedings of the 6th international Conference on Mobile and Ubiquitous Multimedia (MUM '07) vol. 284. ACM, New York, NY, 2007.
16. Storz, O., Friday A., Davies, N., Finney, J., Sas, C., and Sheridan J.: Public ubiquitous computing systems: Lessons from the e-campus display deployments. IEEE Pervasive Computing, 05(3):40–47, 2006.
17. Storz, O., Friday A., and Davies, N.: Supporting content scheduling on situated public displays. Computers & Graphics, 30(5):681–691, 2006.
18. Trimble, J., Wales, R., and Gossweiler, R.: Nasa's MERBboard: An Interactive Collaborative Workplace Platform. In [11], pp. 18–44, 2003.
19. Vajk, T., Coulton, P., Bamford, W., and Edwards, R.: Using a Mobile Phone as a "Wii-like" Controller for Playing Games on a Large Public Display. International Journal of Computer Games Technology, vol. 2008, Hindawi Publishing Corp., 2008
20. Wilmsmann, D., Müller, J., Exeler, J., Buzeck, M., Schmidt, A., Jay, T., Krüger, A.: Display Blindness: The Effect of Expectations on Attention towards Digital Signage. In Pervasive 2009, pp. 1–8, Vol. 5538/2009, Springer Berlin, Heidelberg, 2009.

Audience Measurement for Digital Signage: Exploring the Audience's Perspective

Jörg Müller and Keith Cheverst

Abstract Electronic displays, or digital signs, are rapidly being installed in many public contexts. Increasingly, these signs are equipped with audience measurement, that is, sensors that observe the audience of the signs. However, the audience's experience with audience measurement is not yet well understood, which can be a problem, because the audience may feel observed and that their privacy has been violated. In this paper, we explore the audience's experience with audience measurement for digital signage. We have deployed and evaluated two digital signage systems with different audience measurement techniques, one for a year, and the other for 5 years. We explore the audience's perspective on the employed audience measurement techniques through repertory grid interviews with users who had significant experience as audiences of both deployments.

1 Introduction

As digital display prices drop, paper signs in public spaces are gradually becoming augmented and (in some cases) even replaced by their electronic counterparts. With any media, the measurement on how people react to them, and in particular, how many people look at them (audience measurement), is an important issue, especially when it comes to selling advertising space on these media. Traditional and digital signs are no different. Audience measurement with traditional signage has been performed manually, by counting, observing and interviewing the audience. For digital signs, increasingly automated audience measurement techniques are employed.

J. Müller (✉)
Deutsche Telekom Laboratories, TU Berlin, Ernst-Reuter-Platz 7, 10589 Berlin, Germany
e-mail: hans-joerg.mueller@telekom.de

K. Cheverst
Infolab 21, Lancaster University, South Drive, Lancaster, UK
e-mail: kc@comp.lancs.ac.uk

A. Krüger and T. Kuflik (eds.), *Ubiquitous Display Environments*, Cognitive Technologies, DOI 10.1007/978-3-642-27663-7_9,

© Springer-Verlag Berlin Heidelberg 2012

If digital signage can sense its audience and context, it becomes possible to react to what happens around the sign. For example, the signs can react immediately to the audience (e.g. show a face and have its eyes follow the audience and appear empathetic by appearing to smile when they smile). Additionally, it becomes possible to present *indexical content*, as explained by Kjeldskov and Paay [6], referring to dynamic factors. Symbols (e.g. words) and iconic representations (e.g. images) do not necessarily refer to the immediate environment of the sign. Indices (e.g. '200 m left', 'tomorrow', 'I want YOU for US Army') refer to the spatial or temporal context, or to the audience. While indices to space, time and audience also work for traditional signage, referring to dynamic factors has the risk of the index not working in certain situations (e.g. temporal indices may become outdated).

Audience measurement also more generally simplifies *context adaptivity*. It is important to note that with content shown on traditional or digital signage, the content shown interacts with its surrounding social situation, buildings, expectations, etc. As Mitchell [9], p. 9 puts it: 'Literary theorists sometimes speak of text as if it were disembodied, but of course it isn't; it always shows attached to particular physical objects, in particular spatial contexts, and those contexts-like the contexts of speech-furnish essential components of the meaning.' So, the interpretation of signs is very much dependent on the *place* where they are situated. As Harrison and Dourish [4] emphasize, a place is much more than the mere (constructed) space. It consists of socially constructed patterns, understandings and habits associated with it. Thus, it makes much sense to employ audience measurement to actually adapt the content shown in digital signs to the context.

While audience measurement certainly provides various benefits, there is the possibility that it violates the privacy in the audience. In this paper we explore how the audience experiences audience measurement on digital signage.

2 Related Work

Privacy has been an important topic for all ubicomp technologies, in particular because surveillance is known to have many side effects beyond the mere data collection. For example, it is widely assumed that when people could be observed at any time, but don't know exactly when they are being observed and when not, they are less likely to behave against social conventions. One powerful example is the 'Panopticon' prison design, where guards can observe any prisoner at any time without being seen by them. Classically, studies of the deployed Active Badge system revealed that many users held strong reservations against such 'surveillance technology' [3]. Consequently, attempts were made to design the active badge architecture such that it could provide sufficient user control to mitigate against privacy concerns – however, supporting such user control highlighted a trade off between privacy and efficiency [13]. In order to understand this dimension more thoroughly, we present a set of subdimensions of privacy (from Ref. [8]), as well as a set of guiding principles when designing privacy related systems (from Langheinrich [7]). Both these categorizations are based on the classic work of Altman [1], who defined privacy as 'the selective control of access to the self'. It is emphasized that privacy is neither static nor rule based, but instead a 'dialectic and dynamic boundary regulation process'. Thus, in a dynamic process, users refine the boundary between privacy and publicity according to circumstance. Palen and Dourish [12] describe how this concept applies to computing technology. It is theorized that users continuously try to manage three different boundaries. The disclosure boundary distinguishes the private from the public and needs to be kept in a balance between isolation and crowding. The identity boundary distinguishes the self and the other, where self can also represent a social group one belongs to, and the other or public is also a broadly faced concept with different groups. The temporal boundaries differentiate the past from the present and the future, where it is emphasized that for privacy management the history of interactions as well as the expected future are important. Thus, it is suggested that 'trajectories of action' need to be considered, and that in the balancing process of boundaries small changes may have disproportionately large effects. It is explained that it is less important what technology actually does, but more how it fits into cultural practice and the possibilities of technology are as important as what the technology actually does.

Six subdimensions of privacy are presented in Ref. [8]. Relevant privacy dimensions from this model for audience measurement are:

1. Feedback and Control (Is the user notified and can she switch off the mechanism?)
2. Surveillance versus Transaction (Is the measurement continuous or only when the user performs some action?)
3. Persona versus Activity (Is the audience identified?)
4. Primary versus Incidental content (Does the user want to be measured or does she have a different aim?)

Audience measurement will usually be institutional, and the observer and the observed will usually not be familiar. Therefore the dimensions Interpersonal versus Institutional and familiarity from the model do not directly apply to audience measurement.

Six guiding principles for privacy in Ubicomp are presented in Ref. [7]: Anonymity, Access, Locality, Security, Notice, and Choice. Anonymity as 'the state of being not identifiable within a set of subjects' should be guaranteed where possible. Access describes that data access should be restricted, and data should only be used for purposes that have been specified before the data collection. Locality describes that data should be kept as close as possible to the location where it has been gathered. Adequate Security means that the access to the data should be secured. Notice and Disclosure describes that subjects should always be informed of any data collection, and Choice and Consent states that subjects should always be able to choose whether to participate in the data collection or not. These principles are defined for systems that identify users as well as for purposes of audience measurement. They are certainly applicable and should guide the design of any system.

It should be kept in mind that these dimensions and guidelines are theoretical considerations. Several user studies in the real world indicate that privacy is more an issue of cost-benefit analysis for users. From interviews and practical experiences it is known that many users are willing to trade their privacy for small benefits [15], and people who benefit more from a technology perceive it as less privacy invading [11]. Studies show that although people claim to be concerned about maintaining privacy, in reality they seem often to give away private information easily [2].

3 Exploring Audience Measurement Through Two Digital Signage Deployments

In order to explore the audience's experience of digital signage audience measurement we carried out field tests of five measurement techniques in two deployed signage network prototypes.

3.1 Cameras, Bluetooth Scan and Interaction Logging in the iDisplays *Deployment*

Initially, we decided to deploy the most common measurement techniques, namely: cameras, Bluetooth scanners, and interaction logging, in the iDisplays display network and explore how users would react to these techniques.

iDisplays [10] is a network of 11 public displays (19–42″) installed at Münster University since October 2005, which usually show information about news and events in the university, as well as environment information like weather forecast, cafeteria menu or bus departures. The displays are located in a university institute with 60 regular employees and a few hundred students who visit the institute regularly.

Starting from December 2007, we equipped four of the displays with cameras, small keyboards and Bluetooth scanners. The displays show different alternating content categories such as the information system, videos and comics. All of the audience measurement techniques can be used to sense the sign context as well as the audience reaction.

3.1.1 Face Detection

In order to implement face detection, simple web cams were installed at the displays. The cameras were installed visibly on top of the displays and showed a small red light when capturing video (see Fig. 1). The video was analyzed with a face detection algorithm that would find faces up to a distance of 10 m from the display. When a face disappeared from the camera, the corresponding viewing time

Fig. 1 iDisplay with face detection, keyboard and bluetooth scanner

was stored into a database. No video was recorded or streamed to remote locations, and this was stated to the users when somebody asked.

3.1.2 Keyboards

In order to provide a simple means of interaction with the displays, we installed numeric keypads next to them and these provided a 'Would you like to know more?' functionality. A printout of a screenshot of the display was simply pasted onto the keyboard, such that pressing the item would bring up detailed information on that item. Every time a key was pressed, the event together with the item the user was interested in was stored to a database.

3.1.3 Bluetooth Scanner

Although not very accurate, Bluetooth scanners are often used to re-identify people who have passed by a public display previously. We installed Bluetooth scanners at all displays. Every 11 s the scanners would scan for all visible Bluetooth devices in the immediate environment and store the corresponding MAC addresses to a database. The Bluetooth scanner is a little less well suited to determining audience reaction because of low spatial and temporal resolution. When a certain Bluetooth device is detected, the carrier may already have passed the display, or even be on a different level of the same building. Therefore it is very difficult to attribute Bluetooth scanner data (e.g. a device disappeared) to specific content shown. While the MAC address of the Bluetooth device can be used to recognize devices

Fig. 2 MobiDiC display with coupons and shopfinder

that have been seen before, it may by very difficult to find the person who owns a specific Bluetooth device if he/she does not happen to be in the immediate vicinity.

3.2 *Coupons and Interaction in the MobiDiC Deployment*

In addition to the semi-public iDisplays setting, we were interested in how users react to audience measurement in a 'wild' public display setting. We therefore equipped 20 existing digital signs installed at public telephones in the city centre of Münster, Germany with two different audience measurement techniques as part of the MobiDiC project. In order to measure whether users actually buy an advertised product, coupons for shops in the vicinity are displayed (see Fig. 2) and we check which coupons are redeemed. In order to measure whether users interact with the signs, we record when users download a Shopfinder application for their mobile phone, which will guide the user to the corresponding shop. Both of these techniques are much better suited to determine audience reaction after content was shown than to determine the general context of the sign.

3.2.1 Coupons

We aimed at keeping the couponing system as respectful to the users' privacy as possible. Users can simply take a photo of the screen with their camera enabled mobile phone and show that photo at the corresponding shop. On the display, a code is shown that encodes the current time and the id of the display. The store clerk then records this code, in our case simply by writing it down on a paper form, and the user is given a rebate. The store can then upload the codes via a web form and immediately access statistics about which advertisement worked well in which context. Fig. 2 shows a sample coupon and one of the public telephones.

3.2.2 Shopfinder

As a second measurement technique that measures whether people interact with the displays, a Shopfinder was implemented. When users are interested in a coupon but do not know how to get to the corresponding shop, they can simply send the coupon via Bluetooth to the display. The display then computes the shortest path to the corresponding shop, builds a software guide that shows the way via landmark photos and way instructions, and sends this (in the form of a java application) back to the user's mobile phone. The system ran from September 2007 to September 2008.

4 Use of Repertory Grid Interviews

We were interested in how people would experience the aforementioned techniques after months of real contact, after the novelty effect had worn off. Specifically, we were interested which dimensions that we designed into the systems they considered more important than others and which dimensions would come up that we did not design into the systems at all. In order to explore the dimensions that users used to think about the measurement techniques, we conducted repertory grid interviews [5] after 5 months of deployment.

4.1 Method

Repertory grid interviews have been employed for many decades to elicit the dimensions (constructs) that users use to think about (construe) a certain domain. The benefit of this technique is that the design space is captured purely from the user's perspective, without the interviewer influencing this space through any pre-formed questions.

We interviewed 21 persons who previously had multiple contacts with all measurement techniques, mostly employees of the institute and students. Users were presented the audience measurement techniques in groups of three. For each selection of three techniques, they were asked to state which two of them had something in common that was different from the third technique. They then described what the two had in common (emergent pole) and how the third one was different (implicit pole). These poles could be, for example, 'I can imagine what happens during data processing' versus 'I can't imagine what happens'. After providing a description,

Table 1 Rating for the measurement mechanisms by n = 20 users (Likert Scale where 1 = I like it, 5 = I don't like it)

	Coupons	Keyboard	Shopfinder	Bluetooth scanner	Face detection
Mean	1.7	1.9	2.3	3.6	4
Std. Dev.	0.8	1.0	0.9	1.1	1.3

they were asked to rate each situation on a (five-point) Likert Scale on this dimension. This process was repeated until no more constructs arose. Finally the users were asked to rate each technique on the supplied overall construct of 'I like it' versus 'I don't like it'. The result from each interview with a user was then a number of constructs, together with the rating of the provided situations for each construct. This collection of constructs and ratings provides the repertory grid. To evaluate the grids we applied Honey's content analysis [5], as it enabled us to analyse data from multiple grids (one from each user) and to compare the importance of different constructs by measuring the correlation between the constructs rating and the rating of the overall construct. This correlation (% similarity score) can be categorized for each grid into high (H), intermediate (I) and low (L) correlations, to improve comparability between different grids. The elicited constructs were then categorized in an affinity analysis by two independent raters. Cross-rater reliability of the categories was satisfactory, 59% before and 82% after a harmonization of the categories. For each category a mean % similarity score was computed. This score was then used together with the number of constructs in that category and the HIL values to estimate the relative importance of the given category.

4.2 Results

For the final (supplied) construct in the repertory grid interviews, where one pole was 'I like it' and the other pole was 'I don't like it', coupons were rated best, followed by keyboard interaction, Shopfinder, Bluetooth scanner and finally face detection (see Table 1).

The constructs interviewees used to compare different measurement techniques were grouped into nine different categories (Table 2). The first column states a descriptive name of the category. The second column lists a selection of constructs within this category with highest % similarity scores. For example, user U5 stated that Coupons and keyboards have in common the fact that she can imagine what happens during processing while with the cameras she can't imagine what happens during processing. This rating was 70% similar to her rating for whether she would like the mechanism or not. Thus, this construct was among the top third (H) of all constructs from this user. The next columns state that this category has n = 19 constructs, and that the constructs correlated in mean to 65.26% with the ratings of whether users liked the mechanism or not. Eleven constructs in this category were rated high, while four were rated intermediate and four were rated low.

Table 2 Categories of constructs from the Repertory Grid Analysis in order of importance

Category	Constructs	n	Mean %sim.	H-I-L
1. Feeling of surveillance, fear of misuse, and lack of understanding	**I can imagine what happens during processing** – I can't imagine what happens (70%sim.,H) High potential for misuse – **Low potential for misuse** (70%sim.,H) **Does not intrude my privacy** – Intrudes my privacy (80%sim.,H) **I do not feel observed** – I feel observed (70% sim.H)	19	65.26	11H 4I 4L
2. Choice and effort	**I don't need to use it** – I get utilized (100% sim.,H) **I can choose** – I am at it's mercy (80%sim.,H) **It works automatically** – My active interaction is necessary (50%sim.,H) I have to become active – **I don't need to do anything** (20%sim.,I) I can put it to automatic mode – **I have to become active** (50%sim.,H)	32	54.38	11H 14I 7L
3. Benefit, fairness and feeling of exploitation	Information collection about me – **I want information** (90%sim.,H) **For shopping** – For the display owner (90% sim.,H) **It brings me a benefit** – It doesn't bring me a benefit (100%sim.,H) **It helps in everyday life** – No influence on help (70%sim.,I) **It helps me at my task** – I don't know what it does for me (60%sim.,L)	32	53.26	11H 11I 9L
4. Obviousness and obtrusiveness	**Something obvious** – Hidden, ambiguous (80%sim.,I) **I notice that information is collected about me** – Information can be collected without me noticing it (70%sim.,H)	5	52	2H 1I 2L
5. Perceived anonymity	I can be identified – **Anonymous** (90%sim.,H) **No personal data is collected** – Personal data is collected (70%sim.,I)	12	51.67	1H 6I 5L
Ease of use, usage	**It is immediately clear what I should do with it** – I don't know first what is expected from me (60%sim.,H) I get persuaded – **I have to choose** (50%sim.,I) **I have to touch it** – No touch (60%sim.,I)	18	38.33	3H 7I 8L
Display operator benefit	**I am explicitly interested** – It can also be coincidence (60%sim.,I)	5	30	1H 1I 2L
User hardware necessary?	Hardware is provided – **I get it on my own device** (70%sim.,I)	14	29.29	5I 9L

(continued)

Table 2 (continued)

Category	Constructs	n	Mean %sim.	H-I-L
Miscellaneous	Others can see what I am interested in – **It is done quickly** (30%sim.,L) **I am used to it – It is new for me** (20%sim.,L) **It engages me in a lasting way – It is only for that moment** (10%sim.,L)	8	31.25	2H 1I 5L

The category name, exemplary constructs, number of constructs (n), mean % similarity scores, and High-Intermediate-Low ratings of constructs are mentioned for each category. The pole that correlated higher with whether people like the mechanism is printed bold

In the following, we provide more detail for the five most important categories from the interviews.

4.3 The feeling of surveillance

12 of 21 interviewees mentioned (unprompted) that they would feel monitored by certain measurement techniques, especially the cameras. Unsurprisingly, these constructs had a very high correlation with whether the users liked the measurement technique or not. Many users expressed their feelings as a 'not necessarily rational' awkward feeling when in vicinity of the displays, but also expressed more rational concerns regarding fear of misuse and their lack of understanding. Naturally, these concerns are intensely correlated.

4.3.1 An Awkward Feeling

The mere presence of a camera installed at a display seemed to trigger negative feelings in many users, regardless of the functionality implemented. One user stated,

> Principally I don't have any objections against such a face detection, but I feel a little bit observed by such a camera (U17).

Even more important for display owners, some users experience this feeling as so strong that they assert to even turn away from displays or pass by them more rapidly:

> Somehow, I still look there critically. Because somehow, it feels awkward if I get filmed. I believe that then I look there less. And when I look, I always think, hmm, camera. [...] (U9)

This user explains that the main reason for turning away from the cameras is distrust of statements that the camera image is not captured:

> Even if people say, you never know. There is still the possibility that they capture you. I avoid them as far as I can. But here I cannot avoid them. I just have to pass them. But then I avoid looking there. I hurry past them (U9).

Another user stated that, for her, the effect has worn off over time:

> One is under surveillance and not free in the things one does anymore. But one acts differently if knowing that one gets filmed. In the meantime, I don't care anymore, I just go the ways I have to go. One knows that nobody is sitting behind it (U10).

Another user stated his general feelings like this:

> Because somehow I feel so watched. Even if one says it doesn't get captured. I just don't feel well (U16).

Interestingly, while some users stated they would look less at the displays, there was no obvious behaviour change we could observe.

4.3.2 Fear of Misuse

For some more technology savvy users, the biggest source of fear was less the intended use of the system but more the potential for misuse by criminals or government agencies. Many users expressed serious concerns that a complete infrastructure would be installed for 'harmless' purposes that could easily be turned into a powerful surveillance tool. One user stated

> ... but only for the reason because it enables absolute control. Sure, now it is installed for advertising. But then something changes, and then you just control it. For bilking or whatever (U5).

Another user stated:

> So I pass these cameras every day, and every time I have a queasy feeling. You never know. The computer is connected to the internet; theoretically somebody could crack into it (U15).

The users who had these concerns stated that they were also privately interested in information privacy and thought a lot about this topic.

4.3.3 Lack of Understanding and Trust

Although we communicated that the camera would be used for face detection alone, many users felt unwell because they did not really understand what happens in the software. One user expressed distrust in mechanisms that try to reflect the system activity back to the user, specifically a red dot that shows whether the camera currently captures data:

> Because when I pass the camera, regardless whether there is a red dot or not, I never know what happens with the image: how long will it be saved, what will it be used for in addition to the face detection, will it be saved into a database, you never know (U15).

Other users either did not understand or distrusted the idea that nobody would look at the camera stream:

> Also because I don't know what the camera does, I personally find it a little bit discomforting, when there is a camera in the institute. You don't know, does it capture me, so I don't act stupidly in front of the display then (U22).
> Sometimes when I stand in front of it and see that the camera is on, then I think, oh, what if somebody is sitting behind it (U12).

Another user expressed general distrust about processing the camera stream:

> [With the camera] I don't know what happens, and now I notice how you try the face detection, I find that absolutely suspect (U5).

4.3.4 Sabotage

Especially regarding the cameras, we had various cases of petty sabotage of users who unplugged the cameras or turned them away, e.g. making them point at the ceiling. One camera regularly sabotaged was installed at a display directly above a table at a coffee kitchen, which could capture also a nearby office with a glass door. People who took lunch at the table developed the habit of turning off the screen and unplugging the camera during lunch, but usually restored the state after lunch. People from the office that could be filmed just turned the camera so it would capture the ceiling. Every couple of weeks some random cameras would be turned away and it was not possible to find out who did this.

4.4 The tradeoff between choice and effort

There was a clear distinction between some users who preferred to have the choice whenever any data (although anonymous) would be collected about them and other users who preferred that the displays would adapt to their interests without any effort from their side. One user commented:

> So the camera would get a 5, because I cannot resist at all. That is always the worst, if something is imposed upon me. With Bluetooth, at least I can resist if I disable it (U22).

Another user stated:

> At the cameras I cannot just become invisible, if I don't want to be observed looking at the display. So I prefer it if the user has the choice if one wants to participate (U7).

Other users preferred the cameras, because it would be less effort for them:

> In that case it would be bad for me, that with coupons and shopfinder I have to do something, while at the cameras it is done automatically (U14).

4.5 *The perception of benefit*

The benefit, both monetary and in information, that the user perceives from the audience measurement, clearly had a high correlation with whether users liked the system or not. Users especially seemed to have a very sensible feeling towards the benefit balance, either they would feel that the system would mainly be for their own benefit, or they would feel that the system would mainly be for the benefit of the display owner and they would only be exploited:

> With the Shopfinder I want to have some information [...] And in contrast with the cameras? There I don't want to have anything. I don't want to be watched (U10).

Some users however were not only interested in their own benefit, but also appreciated that the display owner would have a benefit. Asked, whether she liked that any measurement data is collected at all, one user stated:

> Yes, I find that very important for you, so why would you do it if you don't know how well it worked? (U4).

Especially for measurement systems where the user has to participate, the user benefit has to be pretty high. One major reason for disuse of the couponing system for example seemed to be that the experienced benefit was too small for the effort involved in converting the coupons:

> I have photographed it, but I did not go there. For the Gummi Bears. At the train station, spontaneously. I waited at the bus stop anyway. [...] Probably I forgot it then (U14).

4.6 *Hide it or not?*

While some users stated a preference for being able to observe any data collection that takes place, some users stated that they were annoyed by seeing the camera and would prefer it if it would be less visible. While the constructs from the repertory grids indicated that people would prefer the measurement device being obvious, several users stated otherwise in the interviews:

> But because there is the camera and looks like a camera that triggers some process, I believe. I can imagine if I would know that there are some cameras but I cannot see them that I would even prefer it (U18).

Regarding the Bluetooth scanner the user stated:

> I don't see the Bluetooth scanner, I also forgot that it is there, and therefore I almost don't care about it (U18).

4.7 *Perceived anonymity*

The perceived anonymity for different measurement means differed widely. Although it was made clear that the use of the cameras would not store any personal data (i.e. users could not be identified by the stored data), many users still rated the cameras with ‘I can be identified’ or similar constructs. Many users appeared to rate less what data actually were stored, but more the principal possibilities of the device. Some users mentioned that if the device could be limited to only anonymous data at a hardware level (not a software level), they would trust it more:

> So if one could solve this at a hardware level, that would be an advantage. So one would really only give away the number of faces, that would be a pretty anonymous statistic. Then I would find that less bad (U15).

Given guaranteed anonymity, not all users seemed to dislike the cameras:

> So in principle I find the cameras quite funny, but I don’t see yet where that would end. So, I find that pretty good. As long as it stays anonymous somehow (U16).

Interestingly, the Bluetooth scanner was rated as providing a much better anonymity than the cameras with face detection.

5 Design Implications

Audience measurement certainly appears to be a very sensitive topic for many users. In the end, the goal of audience measurement is to collect information about the audience, and audience members may not feel comfortable with this and feel that their boundary between the private and the public is being violated. The interviews indicate that failing to make users trust the displays and feel comfortable in front of them may damage the whole signage system if it causes a significant proportion of passers-by to avoid looking at the displays, hurrying by the displays, campaigning against or even sabotaging the display system.

We adapt the principles introduced by Langheinrich [7] and show how they apply for the specific case of audience measurement. Because privacy is not the only aspect for audience measurement, we propose the additional guidelines of creating a win/win situation, avoiding SPAM, and guaranteeing low effort for the user.

5.1 Anonymity: Guarantee Anonymity by Design

Anonymity, as ‘the state of being not identifiable within a set of subjects’, of the collected data should always be the prerequisite for audience measurement. The fact that a subject cannot be identified by each sensor alone does not, however, guarantee that the collection and amalgamation of different sensor data does also not allow for identification of the individual. In these cases, concepts like K-anonymity [14], which allow the modification of databases in such a way that an individual cannot be identified from the data, should be applied. However, according to our interviews, guaranteeing anonymous data collection is unlikely to be sufficient to win the trust of sceptical users. Despite anonymous data collection, for our users a general feeling of surveillance remained even after months of use. The interviews strengthened what has already been proposed by Palen and Dourish [12]: it is less important what the technology actually does, but more what is principally possible with this technology. Thus, where possible, technology should be applied that technically does not enable more data collection than what is actually used and needed.

5.2 Access and Recourse

In this principle it is stated that proper laws should be defined to forbid unacceptable data collection. Obviously, any system needs to respect relevant privacy laws, however, our interviews indicate that this alone may not be enough to win the trust of users. Many users are very uncertain about what is actually forbidden by law and whether this represents their own concept of privacy.

5.3 Locality

Keeping all data only at the respective display is clearly an interesting idea. Thus, where possible, data should be kept at the respective display, and only where absolutely necessary data should be collected at some central server.

5.4 Security: Design Against Misuse

Security means that any system that processes personal data needs to adequately protect this data from unwanted access, e.g. by encryption. In our case, as the results from the repertory grid and the cases of sabotage indicate, the biggest fear of some users was that the system could be misused for other purposes that those stated, for example by criminals or malicious governments. Indeed, a complete audience measurement network, for example of cameras, could provide significant potential for surveillance that was unimagined by past dictatorial governments or criminals. In order to gain user trust and acceptance, such a system needs to be strongly designed against misuse and users need to perceive this to be the case. Any possibility that the system could be manipulated remotely via a network should be eliminated. One possibility to achieve this would be to implement the complete pre-processing of the raw sensor data in hardware instead of as software on a general-purpose computer. For face detection, for example, the algorithm could run on a dedicated device that only communicates the relevant features, like number of faces, to further processing devices. Once such a design against misuse has been achieved, it needs to be communicated to users to win their trust. With the couponing system for example users felt very comfortable, because it was obvious that no data except the coupon code is transferred from the user to the system, and their was no potential for misuse. With the face detection however users felt very uncomfortable because the processing was done on a PC that was connected to the Internet and was thus very vulnerable to misuse.

5.5 *Notice: Create Transparency but Don't Be Obtrusive*

Notice and Disclosure means that no single personal data collection can go unnoticed of the subject that is being monitored. One prerequisite to gain the trust of users should be to have them understand what the system does. In our case, the dimensions of surveillance, obviousness and anonymity from the repertory grids and the cases of sabotage indicate that there was a general lack of understanding of how the measurement systems work. However, for explanations of the system or reflections of the system state to the user, like the red camera light, users seem not always to trust them. In our case, creating transparency seems to pose a dilemma: While many users stated that they would distrust a system where they could not clearly see the sensor hardware, some users also expressed that they do not like to be reminded of the data collection all the time, for example by a visible camera. It could be best to install the device in an unobtrusive manner, like a camera behind darkened glass, but make the data collection transparent, like with certain icons that express what data are collected. With the coupons and the keyboards, for example, how the system worked was very transparent for users as was how the data was processed, while for the other systems transparency was much less.

5.6 *Choice: Give Users a Choice but Make It Little Effort*

Regarding the principle of choice and consent, Langheinrich [7] explains that any personal data collection requires the explicit consent from the user. Again, for anonymous data collection, this is not necessarily the case. In our case, as the dimension of choice and effort from the repertory grid indicates, many users prefer measurement systems where they can actively decide whether they want to participate or not, like the coupons, Shopfinder or keyboard interaction. Systems that were imposed on the users, like the Bluetooth scanners or even more so the face detection, were experienced much less favourably. We suggest that systems should be designed so users can opt out. Were this is not possible, at least the sign should be installed in such a way that users can avoid it, thereby opting out. Installing offensive audience measurement in a place where users cannot escape, such as in an elevator, is probably not a good idea.

In addition to these adapted privacy guidelines, we propose a number of specific guidelines for digital signage audience measurement.

5.7 *Create a Fair Win/Win Situation*

The benefit to the user is not mentioned by Langheinrich, but played a clear role in our case. Taylor [15] explains that many users are willing to sometimes trade their privacy if the benefit is high enough. In our case, for most users it seemed to be important that the measurement was experienced as a fair win/win situation (as with the coupons), where they don't feel exploited (as with the cameras).

Value for the audience can be considered either in terms of information gain (as with the Shopfinder or the keyboard interaction) or in monetary terms (as with the Coupons). As the low uptake of the coupons indicates, especially when users have to actively participate, a strong incentive is necessary.

5.8 *Be Careful Not to SPAM*

For systems that require the participation of the audience, the measurement itself could clearly be experienced as annoying by users. When we developed the couponing mechanism, we considered distributing the coupons to any visible Bluetooth device in the environment. This was perceived very negatively and annoying by many users. The pull mechanism of having users take a photo of the display was perceived much more favourably.

Besides the signs themselves respecting the user's attention, designers should also take care that the means for carrying out the measurement does not itself annoy users.

5.9 *Make the Measurement Mechanism Low Effort for the User*

Comparing the different mechanisms, there is a clear correlation that the less effort the mechanism posed for the user, the more data could be collected. Because the benefit users gain from audience measurement will usually be very little, even a very small effort required from the user, like taking out the mobile phone, may significantly reduce data collected.

6 Conclusion

In this paper we have explored the issues that arise from audience measurement of digital signage. Our exploration has focussed on the audience's perspective and, crucially, has utilised data (gathered through repertory grid interviews) from users who have actually experienced a variety of audience measurement techniques.

Based on our analysis of this data we have derived and presented a set of 'implications for design' which should be of value to those wishing to design and develop digital signage systems which have the capability to react and respond to their audience without causing a significant proportion of their potential audience to deliberately avoid engagement with the digital signage.

References

1. Altman, I.: The Environment and Social Behavior: Privacy, Personal Space, Territory and Crowding. Brooks/Cole Pub. Co. (1975)
2. Berendt, B., Günther, O. and Spiekermann S.: Privacy in E-Commerce: Stated Preferences vs. Actual Behavior. In: Communications of the ACM vol. 48 (4), pp. 101–106 (2005)
3. Harper, R.: Why People Do and Don't Wear Active Badges: A Case Study, In: Computer Supported Cooperative Work, vol. 4 (4), pp. 297–318 (1995)
4. Harrison, S. and Dourish, P.: Re-place-ing space: the roles of place and space in collaborative systems. In: Proceedings of CSCW'96, pp. 67–76 ACM (1996)
5. Jankowicz, D.: The Easy Guide to Repertory Grids. Wiley (2003)
6. Kjeldskov, J. and Paay, J.: Indexical interaction design for context aware mobile computer systems. In: Proceedings of OZCHI'06, pp. 71–78. ACM (2006)
7. Langheinrich M.: Personal Privacy in Ubiquitous Computing. PhD Thesis. ETH Zürich (2005)
8. Lederer, S., Mankoff, J. and Dey A.: Towards a Deconstruction of the Privacy Space. In: Proceedings of the Ubicomp 2003 Workshop on Ubicomp Communities (2003)
9. Mitchell, W. J.: Placing Words. Symbols, Space, and the City. MIT Press (2005).
10. Müller, J., Paczkowski, O. and Krüger, A.: Situated Public News and Reminder Displays. In: Proceedings of European Conference on Ambient Intelligence 2007, pp. 248–265. Springer (2007)
11. Newman, W., Eldridge, M. and Lamming, M.: PEPYS: generating autobiographies by automatic tracking In: Proceedings of CSCW 1991, pp. 175–188. Kluwer (1991)
12. Palen, L. and Dourish P.: Unpacking "Privacy" for a Networked World. In: Proceedings of CHI 2003, pp. 129–136. ACM (2003)
13. Spreitzer, M. and Theimer, M.: Scalable, Secure, Mobile Computing with Location Information. In: Communications of the ACM, vol. 36 (7), p. 27. ACM (1993)
14. Sweeney, L.: Achieving k-anonymity privacy protection using generalization and suppression. In: International Journal on Uncertainty, Fuzziness and Knowledge-based Systems, vol. 10 (5) pp. 571–588 (2002)
15. Taylor H.: Most People Are "Privacy Pragmatists" Who, While concerned about Privacy, Will Sometimes Trade it off for other Benefits. The Harris Poll (2003)

Analysis and Prediction of Museum Visitors' Behavioral Pattern Types

Tsvi Kuflik, Zvi Boger, and Massimo Zancanaro

Abstract Personalization in the "museum visit" scenario is extremely challenging, especially since in many cases visitors come to the museum for the first time, and it may be the last time in their life. There is therefore a need to generate an effective user model quickly without any prior knowledge. Furthermore, the initial definition of a user model is also challenging since it should be built in a non-intrusive manner. Understanding visitors' behavioral patterns may help in initializing their user models and supporting them better. This chapter reports three stages of analysis of behavior patterns of museum visitors. The first step assesses, following past ethnographic research, whether a distinct stereotype of behavior can be identified; the second shows that visitors' behavior is not always consistent; the third shows that, in spite of the inconsistency, prediction of visitor type, is possible.

1 Introduction

The museum environment is both an attractive and a challenging arena for developing and experimenting with personalized information delivery. It is rich in information, offering much more than a visitor can experience in a museum visit of limited time, and hence personalization may be a tool for enhancing the visit experience and help the visitor avoid information overload. The need for personalization is also motivated by

T. Kuflik (✉)
The University of Haifa, Mount Carmel, Haifa 31905, Israel
e-mail: tsvikak@is.haifa.ac.il

Z. Boger
Ben-Gurion University of the Negev, Beer-Sheva 84105, Israel
e-mail: zboger@bgu.ac.il

M. Zancanaro
FBK, via Sommarive 18, Povo 38050, Italy
e-mail: zancana@fbk.eu

A. Krüger and T. Kuflik (eds.), *Ubiquitous Display Environments*,
Cognitive Technologies, DOI 10.1007/978-3-642-27663-7_10,
© Springer-Verlag Berlin Heidelberg 2012

recent research regarding museum visitors' identity [7]. However, to achieve this goal, an accurate model of the user must be built quickly in order to provide a personalized service at the earliest stage possible of the visit.

In this work, we adopted the classification of visiting styles proposed by the ethno methodologists Veron and Levasseur [20]. Starting from ethnographic observations of the behavior of a number of visitors in the Louvre museum, they argued that visitors' movements may be classified into four categories that, for the sake of clarity, they named after the "typical" behavior of four "typical" animals. Specifically, they suggest that:

1. The *ANT* visitor tends to follow a specific path and spends a lot of time observing almost all the exhibits;
2. The *FISH* visitor most of the time moves around in the centre of the room and usually avoids looking at exhibits' details;
3. The *BUTTERFLY* visitor does not follow a specific path but rather is guided by the physical orientation of the exhibits and stops frequently to look for more information; and finally,
4. The *GRASSHOPPER* visitor seems to have a specific preference for some pre-selected exhibits and spends a lot of time observing them while tending to ignore the others.

Of course, it might be expected that a given visitor can change her behavior during a long visit, and it is also possible that the style is affected by the specific interests.

In this work, we started from the very basics: we empirically validated Veron and Levasseur's model of visiting style. We then looked at the consistency of the visitors' exhibited behavior during the visit, and finally we tried to identify the visitors' type based on initial observations at early stages of the visit. We used log files of 140 visitors exploring a frescoed room with a multimedia museum guide, first to provide quantitative-based evidence that museum visitors' behavior may effectively be classified according to Veron and Levasseur's mode and then to analyze their behavior and to try to identify their type. This work is intended to complement and build on Veron and Levasseur's ethnographic study by providing empirical evidence, as well as to provide information in a principled way for further research on user modeling, and then to extend it for automatic identification of visitors' types at the very beginning of the visit.

2 Related Work

Many studies have investigated personalized information presentation in the context of mobile museum guides, as surveyed by Baus and Kray [2]. Several research prototypes focused on user preferences, initial knowledge and history of interaction. For example, the GUIDE system presented in Cheverst et al. [5], adapted web-like presentations by adding information about nearby attractions that might interest the visitor to a city. The HIPPIE system proposed personalized tours in a museum by maintaining a model of user interests and knowledge [14]. The REAL system [1] adapted route descriptions according to the actual user position, the limited technical resources of the device, and the cognitive resources of the user. In the context of the PEACH project [18] a spreading activation technique applied on a domain knowledge-base was implemented to predict the interest in concepts related to those for which the system received explicit feedback from the user [9, 10].

Knowledge and interest-related features are not, however, the only sources of information that are worth considering for modeling a museum visitor. For example, Petrelli and Not [15] suggested considering whether the user is visiting the museum alone or with companions, whether it is a first-time or a recurrent visit, and so on.

Behavioral traits have also been taken into consideration. Sparacino [16] proposed categorization of user types into three main categories: (1) the greedy visitor who wants to know and see as much as possible; (2) the selective visitor who spends time on artifacts that represent certain concepts only and neglects the others; and (3) the busy visitor who prefers strolling through the museum in order to get a general idea of the exhibition without spending much time on any exhibits. Her application employs Bayesian networks to model both the user (interest and style) and the appropriateness of the guide's content (length and order).

The same categorization of user types is also used by Hatala and Wakkary [8] together with an ontology-based model of the interests. In both these papers, the validity of such a scheme is justified through qualitative analysis, mainly site studies and interviews with staff at various museums.

In addition to research into exploiting the potential of novel technology in the museum, recent research studies in museums point out the need for personalized support for visitors. Falk [7], based on a lifetime of research in the area, concluded that the notion of *identity* is important for understanding the reasons that bring a person to a museum and defining their goals. He identifies five major categories of identity-related groups: Explorer, Facilitator, Experience Seeker, Professional/ Hobbyist, and Recharger. He emphasizes that whatever the visitor sees or does is influenced by the combination of their identity, his or her personal context constituting prior knowledge, experience and interest, the physical context, that is, the specifics of the exhibition he/she encounters, and the socio-cultural context, that is, the within- and between-group interaction that occurs in the museum. Hence, the visitor perceives the museum experience as satisfying if there is a good match between the visitor's identity-related needs and the museum affordances. During and just after the visit, the visitor constructs meaning from the experience, where this is particularly related to identity building. If a system is to improve the visitor's experience, it should be able to do this better if it takes into account the visitors' type, their personal context that is composed of prior knowledge, experience and interests, social context, and the museum's physical context. Hence personalization in museums is not just about exploring novel technology capabilities but also answering a real need.

Concerning the Veron and Levasseur classification, the first attempt to exploit it as part of a user model for a mobile guide was in the HIPS project (mainly Ref. [13]). They used a Recurrent Artificial Neural Network (ANN) trained to recognize the visiting style of a visitor given her interaction history. Although most of the ideas that were tested experimentally in HIPS underwent user evaluation, the very idea of the existence of visiting styles was taken for granted, relying on the qualitative analysis of the original work.

Chittarro and Ieronutti [6] employed Veron and Levasseur's classification in the context of a tool that visualizes users' behaviors in a virtual environment. Their use of the visiting styles was based on qualitative analysis. They did not evaluate the existence of these classes and they did not use it for personalization.

The use of behavioral traits for user modeling has also been taken into consideration by Sparacino [16], although she used different categories. Her application employs Bayesian networks to model both the user (interest and style) and the appropriateness of the guide's content (length and order).

The same categorization of user types was also used by Hatala and Wakkary [8] together with an ontology-based model of the interests. In both these works, the validity of such a scheme is justified through qualitative analysis, mainly site studies and interviews with the staff at various museums.

Recently, Bohnert et al. [4] tried to predict visitors' behavior automatically by using manually annotated visitors logs. They tracked and annotated visitors' paths through a museum exhibition, recording where they stopped and for how long and then tried to train a variety of classifiers, with different features to predict visitors' behavior. They tried to predict the visitors' next K stops, once based on their interest in already seen exhibits, and then again based on the visitor's path similarity to that of other visitors through the museum. They also experimented with a combination of the two (hence an interest-based classifier, a collaborative classifier, and a hybrid one).

3 Research Environment and Experimental Results

In the context of a user study of a multimedia mobile guide [17], 143 regular visitors to Torre Aquila[1] in Trento were invited to test the system. Among the subjects, 61 were males and 82 females. Their age ranged from 20 to 79 years (mean = 47, median = 50, std. dev = 15.9). The visitors were recruited at the entrance of the museum and got a free ticket to visit the castle as a reward for participating in the experiment. The visitors used a mobile multimedia museum visitors' guide and their interactions with the system, such as presentation selection and giving

[1] Torre Aquila is a tower at the Buonconsiglio Castle in Trento, Italy, where a fresco called "The Cycle of the Months," a masterpiece of the Gothic period, is displayed. This fresco, painted in the Fifteenth Century, covers all fours walls of a room in the tower and illustrates the activities of aristocrats and peasants throughout the year. The museum guide used to collect visitor data is one of the many prototypes developed in the PEACH project. For more details see Ref. [17].

feedback, were logged throughout the visit using time-tagged Infrared-based positioning information. Out of the 143 visit logs, 140 were used for this study; the rest had various errors that prevented their use.

3.1 Identifying Distinct Visitors' Stereotypical Groups

In the first stage we were interested in trying to identify four distinct groups of visitors based on their behavioral patterns.[2] Hence we used measures that related to the entire visit rather than temporal-based indices that may be better suited for prediction. The measures used for the analysis were the following: *AvT*, average time a visitor spent at each position; the percentage of exhibits visited during the visit (a binary variable where 1 means the visitors spent some time at every exhibit and 0 means the visitor did not stop at any exhibit); *Order*, a numerical representation between 0 and 1 of the order of the visit, where 1 means that the visitor walked from one exhibit to the other, following the natural order of the exhibition, while 0 means that the visitor jumped back and forth in unordered manner; *Completeness*, a combined description of visitors' interaction with the museum visitors' guide system, taking into account interaction and whether or not visitors listened fully through complete presentations. Further, four cumulative measures, represented by the four letters A, B, C, D, were defined that considered the percentage of the visit for which the visitor was: *A*, interacting with the guide, i.e., giving feedback and asking for more information, but not reaching the conclusion at the end of the presentations; *B*, interacting and reaching conclusions; *C*, not interacting and not reaching conclusions; and *D*, not interacting but reaching conclusions. The sum of A, B, C and D is 100%; each part presents the percentage of the visit where the visitor exhibited the type of behavior represented by the specified letter.

Data pre-processing generated 140 seven-dimensional vectors including the average time, visit order and completeness, and the percentage of the visit for which the visitor's behavior was according to each of the four types represented by the letters. Checking for each and every position whether the visitor interacted with the system or not and whether s/he viewed complete presentations or not and then calculating the ratios.

Two different clustering approaches were used independently, Artificial Neural Networks (ANN) and K-means algorithm (MacQueen 1967), and their results were compared.

Artificial neural networks are used to form data-driven models. In order to perform unsupervised learning, an auto-associative ANN (AA-ANN), in which the "targets" are identical to the inputs, was used. If the trained AA-ANN succeeds in replicating the inputs as outputs, it means that the hidden neurons are encoding

[2] It is worth noting that here we report only the main findings, since this part is also reported in detail by Zancanaro et al. [21].

Table 1 Confusion matrix for the classifications based on the ANN and K-means clustering

ANN labels × K-mean labels crosstabulation						
Count		K-mean labels				Total
		A	B	F	G	
ANN labels	A	50	1	0	2	53
	B	1	33	2	0	36
	F	0	2	12	1	15
	G	2	0	3	31	36
Total		53	36	17	34	140

the essential information "distilled" from the inputs features. In most cases the outputs of the hidden neurons are close to either one or zero [3]. Thus all examples that generate the same hidden neurons output pattern are deemed to belong to the same cluster.

The AA-ANN used was a fully-connected, feed-forward ANN of two hidden neurons and seven output neurons, each having the sigmoidal transfer function, which was presented with the dataset with the seven input variables and the identical values as targets.

K-means is a well-known technique of clustering data-points by maximizing the distances among the clusters of similar data points.

We clustered the seven-dimensional vectors into four clusters, using the two different methods. The clusters were then manually annotated using the "animals" characteristics. It was possible to do this annotation for both AAN and K-means clusters.

In order to assess to what extent the two clustering algorithms agree on classification of the visitors into the different visitors styles, we used the κ statistics [12], which provides a better estimation of the bare percentage agreement since it takes into account the possibility of chance agreement. Table 1 shows the confusion matrix. The value of the κ statistics in our case is 0.860 with a standard error of 0.035 ($p < 0.0001$; $N = 140$). According to Landis and Koch's criteria [12], the agreement is very good ($\kappa > 0.8$).

The results of the first study showed that the two algorithms, which are based on different principles, provided similar very results and classified the data along the same characteristics that were close to the ones predicted by the Veron-Levasseur model. Therefore, we can assume that the four "animals" are a natural way of categorizing visitors when the variables above are chosen.

3.2 *Analyzing the Consistency of Visitors' Behavior*

The results of the first study paved the way for the second one, which was aimed at exploring how consistent the visitors were in their behavior. The measures used for the analysis were the same seven measures used for the previous study and a few more. At each position, we measured the portion of the time dedicated to viewing a

Table 2 Duration of the visit according to visitor types: the differences between all types were found significant (two-tailed *T*-test, $p < 0.05$) with the exception of the Fish and the Grasshopper

Type	Average	Ant	Butterfly	Fish	Grasshopper
Av. time	25.18	29.7	24.3	19.1	21.2
STDEV		4.3	3.7	4.0	6.7
Number of cases		56	37	12	35

presentation as well as the portion of "idle" time, that is, time spent at a position when no presentation is played. We also noted start time and end time. In total, the dataset was composed of 140 11-dimensional vectors. In order to explore behavioral changes over time, the log files were sampled every 2 min in order to generate vectors representing the cumulative parts of the visit in 2-min intervals. The first set of vectors represented the first 2 min of the visit, then the second set of vectors represented the first 4 min of the visit, etc. Hence we had sets of 11-dimensional vectors, representing, for every visitor, "cumulative visit behavior" in 2-min intervals, from the beginning of the visit until its end. The 2-min time interval was selected as it seemed to be most appropriate to the task: we had six positions and people tend to stay about 3–4 min at every position, so this gave us between one and two measures at every position.

For this experiment we used only the ANN. To simulate a real-time application, in which a new visitor behavior type is learned on-line, an AA-ANN model was trained repeatedly every 2 min, using the accumulated information up to that point in time. The AA-ANN architecture had 11 features, both as inputs and targets, and 2 hidden neurons. The trained ANNs were used to categorize the partial vectors every 2 min and thus, for every visitor we got a series of categories from the time he/she started the visit until they left the exhibition. Ideally, if the visitors exhibit consistent behavior, we can expect that the categorization of the partial vectors will be the same as the final categorization, as represented by the AA-ANN. It is worth noting that in this step we were interested in assessing the characteristics of the behavior of the individual visitors and not the predictive capability of the ANN, and hence we performed training, using the whole data set, without any testing stage.

Since visit durations of the various types differ (see Table 2 for details), and in order to be able to analyze and compare the results, we converted the time-based results ("snapshots taken every 2 min") to results that are relative to the visit length. We calculated the categorization results after the first 10% of the visit, after 20%, etc. Table 2 presents, for the different visitor types, the average time of a visit and the standard deviation of it and the number of visitors for every type.

In order to evaluate the categorization, we calculated the accuracy as ((TP + TN)/ (TP + TN + FP + FN)) of the categorization every 10% of the visit time where TP stands for true positive, TN stands for true negative, FP stands for false positive FN stands for false negative. Figure 1 presents the accuracy of the categorization results: on the X axis there is the portion of the visit (1 = end of visit) while on the Y axis is the accuracy.

Looking at Fig. 1, we can see that the average accuracy starts at a level of about 55% and increases linearly to 90% at the end of the visit. Therefore, it is possible to

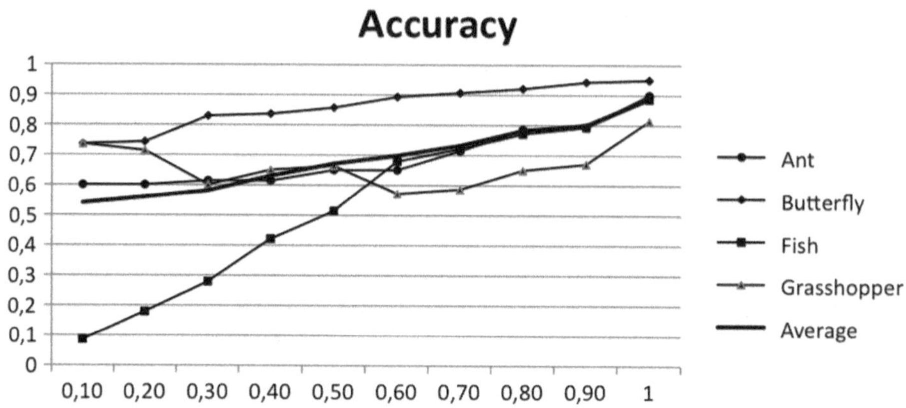

Fig. 1 Accuracy of predicting museum visitors' type based on their behavior during the visit

predict the visitor type at an early stage of the visit and we may be able to produce better estimations as the visit progresses. Yet, looking at the individual visitor types, some are better estimated than others. For example, the accuracy in predicting *BUTTERFLY* is very good from the very beginning. *ANT* prediction is quite good at the beginning and improves with time. *FISH* is not good at the beginning but constantly and quickly improves (after about 50% of the visit it is like all others) and *GRASSHOPPER* prediction seems to be good at the beginning (over 70%) deteriorates a bit to about 57%, and then improves again.

We should note that the accuracy is composed of correct acceptance of a right categorization and correct rejection of a wrong categorization. A low level of false positive is good, so we do not make mistakes. However, we also need good prediction (high level of true positive). Hence we took a closer look at the quality of the positive prediction, known as "precision" measure in information retrieval (TP/(TP + FP)) and the comprehensiveness of the prediction, known as "Recall" (TP/(TP + FN)). These two measures are used in combination to present the effectiveness of search engines: how "clean" the results returned to the user are and how comprehensive the returned set is. Both represent two aspects of how well the search engine finds the relevant information available. Values of precision and recall vary between 0 and 1; the higher the value, the better the performance. We calculated the Precision of the categorization that shows how accurate the practical categorization is, i.e., when the system categorizes a visitor, how accurate this definition is. We calculated also the Recall, to show how well the system can find all users that belong to a specific category.

Figure 2 presents the precision of the prediction of the visitors' type. The X axis represents the portion of the visit (1 = end of visit) while the Y axis represents the precision. As can be seen in Fig. 2, quite early during the visit the system is able to identify an *ANT* and a *BUTTERFLY* accurately and maintain this good identification throughout the visit. The practical meaning is that once a visitor has been

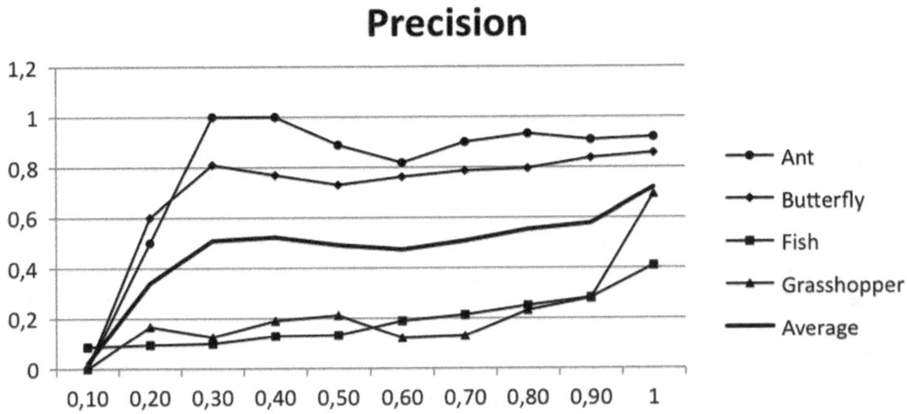

Fig. 2 Precision of predicting museum visitors' type based on their behavior during the visit

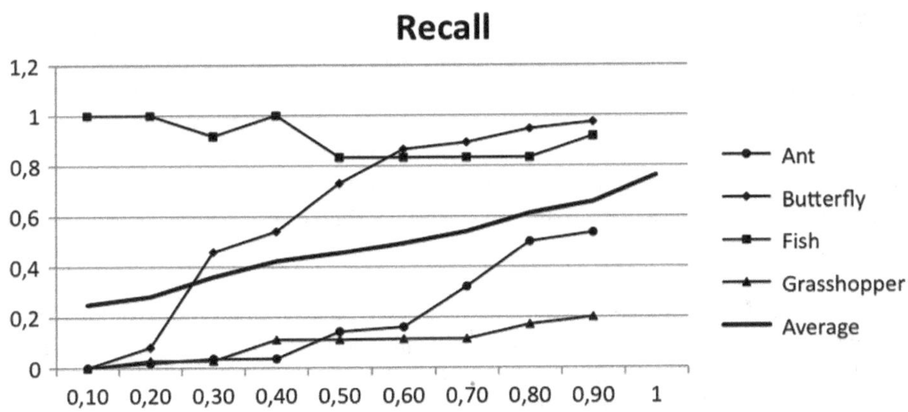

Fig. 3 Recall of predicting museum visitors' type based on behavior during the visit

identified as an "*ANT*" or a *BUTTERFLY*, the identification is good, while the *FISH* and *GRASSHOPPER* are not identified successfully until very late during the visit.

Figure 3 presents the recall of the prediction of the visitors' type as a function of the progress of the visit. It can be seen in Fig. 3 that the system identifies all, or most, of the visitors of type *FISH*. Many visitors of type *BUTTERFLY* are identified properly sometimes, and their identification improves over time. The visitors of type *ANT* tend to be identified relatively late during the visit while visitors of type *GRASSHOPPER* are generally missed.

As can be see in Figs. 1, 2 and 3, while the categorization in general is quite good – accuracy of over 0.5 at the very beginning and improving as the visit evolves, there are major differences between the various types. The accuracy of categorization predication for the *BUTTERFLY* seems to be extremely good, for the *ANT* and *GRASSHOPPER* quite good, and for the *FISH* poor at the beginning, but

Table 3 Visit types stabilization time

	Average	Ant	Butterfly	Fish	Grasshopper
Time	15.1	21.6	9.0	0.9	15.4
STDEV	8.0	4.9	4.4	2.7	4.6
Percentage	58%	72%	38%	4%	75%

All differences are found to be significant (two tailed *T*-test, $p < 0.05$)

improving dramatically towards the middle of the visit. Looking at Figs. 2 (precision) and 3 (recall), if an *ANT* or a *BUTTERFLY* is identified, the identification is accurate, while identification of a *GRASSHOPPER* is not good and identification of a *FISH* is poor.

The differences between the accuracy and the precision/recall of the prediction lead us to examine further the visitors' behavior, since it can be expected that visitors' behavior will change along the visit (and that the measured data will reflect that).

Indeed, we found out that at the very beginning, all visitors are first predicted to be *FISH* which is a reasonable expectation since the *FISH* is the category with little need of information. This means that the data accumulated at the first 2-min interval of the visit are not enough for visit style prediction. Moreover, on the average, during the first 10 min many visitors were defined as *FISH*. The prediction tends to change and stabilize on the correct type after about 9–15 min. Table 3 shows the average "stabilization" time (we define stabilization as the convergence to the accurate type, as was defined manually) and percentage of the visit for the various types.

Table 3 emphasize that the *ANT* behavior requires quite a long learning time, but once learned, the identification is accurate, as presented by the precision. *BUTTERFLY* behavior is identified more quickly, while identification of *FISH* may be immediate, but the fact that with little information almost all visitors are identified as *FISH* explains the low precision. The *GRASSHOPPER* seems to be the trickiest to identify.

The results of the analysis of the consistency of visitors' behavior were a bit discouraging. It seemed that, as may be expected, visitors do not exhibit well defined and consistent behavior throughout the visit.

3.3 *Results: Predicting Museum Visitors' Visiting Patterns*

In spite of the discouraging results of the second step, we wanted to evaluate the system's ability to predict visitor type at early stages of the visit. Given the relatively poor results of the previous stage, we decided not to rely on single cumulative data points, as we did before, but to use a growing number of samples and to see if the predication capability improved as the visit progressed. We used vectors that are multiplications of the original vectors – 11-dimension vectors represented the first 2 min, 22-dimension vectors represented the first 4 min (where the first 11-dimension represented the first 2 min and the second

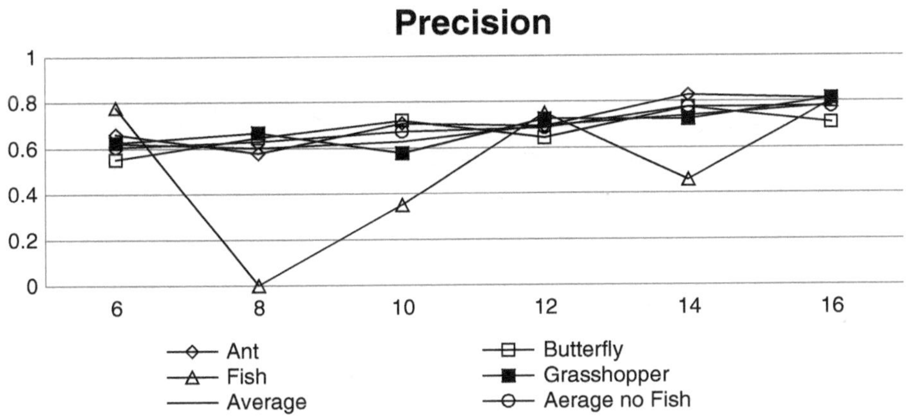

Fig. 4 Precision of the prediction of visitors' type as a function of time into the visit

11-dimensions represented the first 4 min – including the first two), so every vector represented the evolution of the behavior of the visitors. Again, we used the ANN for the study; however, this time, since we aimed at identifying visitor types, we performed tenfold validation; we trained the ANN with 90% of the data and tested it with the 10% that were not used for testing. This process was repeated ten times. We performed the study for vectors representing visits after 6 min from the beginning of the visit, since before that we do not have enough data, until 16 min from the beginning of the visit, since at that time some of the visitors started to leave the museum and we wanted to be able to use data representing realistic scenarios.

For analyzing the accuracy of the prediction of visitors' type we used again precision and recall. Figure 4 presents the results for visitor type prediction precision. It presents the average results of the ten repetitions, for the individual visitor type and a general average. Additionally it presents the average of *AN*", *BUTTERFLY* and *GRASSHOPPER* (to be discussed shortly). Looking at Fig. 4, it seems that, in general, prediction gradually improves from a little over 0.6 at 6 min into the visit to 0.8 after 16 min of the visit. While 0.6 may seem a bit low, 0.8 seems quite promising.

Turning to the individual graphs, we note that, while all exhibit gradual improvement, the *FISH* type exhibit a strange behavior: a missing value at 8 min and deteriorating performance at 12 min. Since this behavior seems problematic (and will be discussed later), we presented also an average of results without taking into account the *FISH* type, which now shows a slight improvement.

As already discussed, while precision represents the accuracy of the performance, i.e. the proportion correctly classified by the system, this is only half of the picture. Another question is whether the system is able to find all, or almost all, items of a given class. Hence we need to look at recall as well. Figure 5 presents the recall of the prediction. Like Fig. 4, it presents the recall for the individual types as well as the two averages. Looking at Fig. 5 we see a similar behavior: gradual

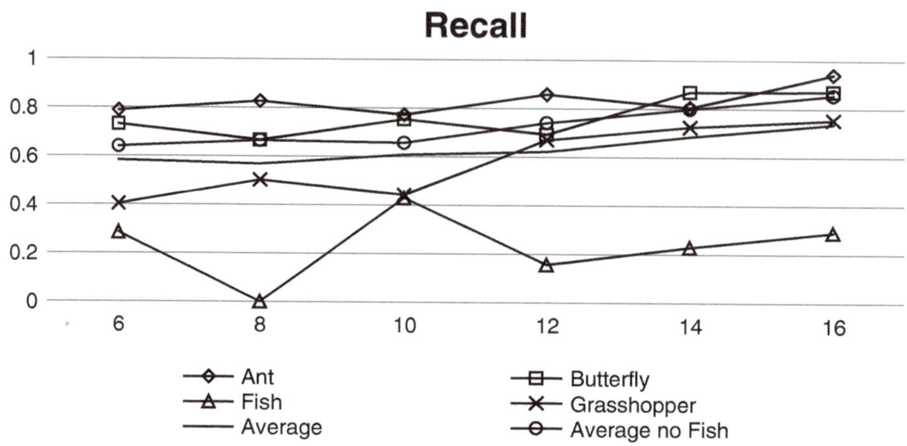

Fig. 5 Recall of the prediction of visitors' type as a function of time into the visit

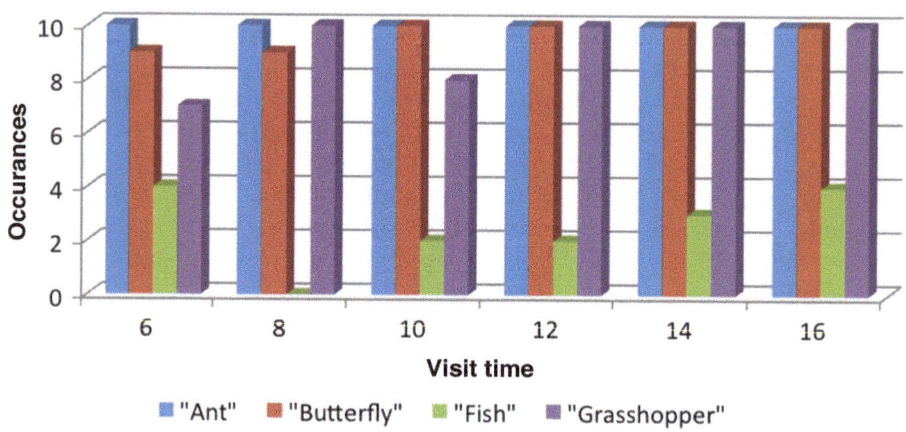

Fig. 6 Level of success in identifying visitors' type at every point in time

improvement in average, from a little below 0.6 to a little above 0.7. Again, *FISH* behavior is quite strange and taking this type out of the consideration results in a noticeable improvement from a little above 0.6 to about 0.85. Looking at the individual graphs we notice that the system is able to find more *ANT* and *BUTTERFLY* types than *GRASSHOPPER* at early stages, but the difference is getting smaller as time passes. Overall, we can see that the system is able to categorize the visitors correctly.

In order to understand this behavior better, we looked into the ten individual runs and checked if the ANN was able to identify *FISH* every time. Figure 6 presents the ability of the ANN to identify visitor types at every individual run. In Fig. 6 we clearly see that the system fails to identify *FISH*. At the first step it succeeds in four out of ten attempts, at the second step (8 min) it fails completely; this explains the "hole" in the results.

Looking back at Table 1, we see that in general there were 12 visitors classified as *FISH* on which the two clustering algorithms agreed (out of 17 or 15, based on the system and 14 based on manual annotation). Hence it is obvious that the system did not have enough training data and that for this case of analysis *FISH* is not predictable. It seems that even though visitors' behavior tends to be inconsistent, as more data are accumulated it is possible to predict visitors' types.

4 Discussion

This work follows and extends our previous work [20], which showed that it may be possible to identify different generic types of museum visitors, based on their behavior. The ability to distinguish types of visitors may lead to services that are better adapted to museum visitors, a requirement shown by recent museum research to be important. The aim of the work was to validate the finding of the ethnographical research and to examine the possibility of predicting visitors' behavioral types correctly, considering that visitors may behave inconsistently. We were able to identify four distinct types characterized by distinct behaviors. We also found that the behavior of the visitors tends to be somewhat inconsistent and it may take time until the type "stabilizes", i.e., converges to a consistent animal type. However, in spite of this inconsistency, we were able to identify three out of the four visitor types accurately, while the fourth type was probably not identified simply due to lack of data. Once visitors are identified, in the early stage of the visit, this may allow us to adapt services provided to the visitors according to the visitor's type to which they belong. Information delivered to an *ANT* may be detailed and thorough while information delivered to a *FISH* may be of an overview type. Information delivered to a *GRASSHOPPER* may be detailed, but limited to the specific objects of interest, while the information delivered to a *BUTTERFLY* may be "in-between" *FISH* and *GRASSHOPPER*. Our results show that a visitor's type can be accurately identified rather early during the visit, within 10–20 min, assuming a museum visit may take a few hours. Hence, once identified, these visitors may receive information tailored to their "type": detailed and thorough information for the *ANT*; more generic an overview for the *BUTTERFLY*. A *GRASSHOPPER* seems to be a little less easily identifiable. Identification of the *FISH* requires a bit more research.

As noted earlier, the overall Accuracy of the results seemed encouraging, but a closer look revealed differences between the different visitors' types. This was better analyzed using the classical complementary measures of Precision and Recall. However, in the museum scenario, we may not treat precision and recall equally. While we would like to be able to categorize all visitors accurately as early as possible, reality may be different. This raises the question of what is more important, precision, that is the correct categorization of visitors, or recall, that is, comprehensive categorization of as many visitors as possible. It seems to us that in our case it is more important to categorize visitors accurately than to try to categorize all of them while making errors in the process. We had better miss an

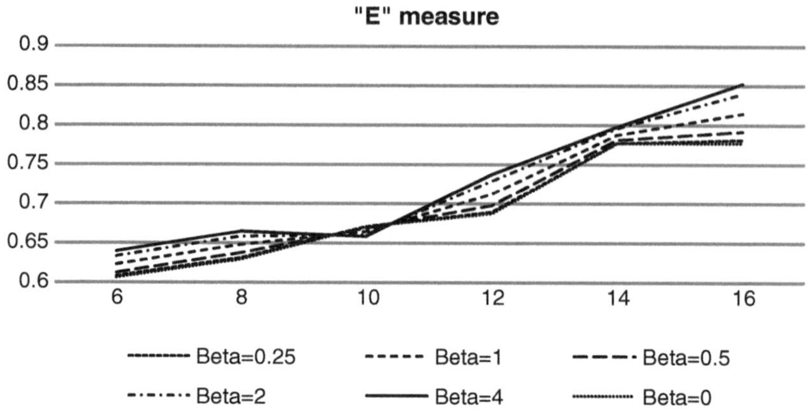

Fig. 7 "E" measure with different B values

ANT, and provide no service, than treat an *ANT* as a *FISH*, and provide the wrong service. In practice, such errors may cause the system to exhibit undesired behavior, such as presenting short and general presentations, intended for the *FISH* type of visitor to an *ANT* type of visitor identified as a *FISH* at the early stage of the visit.

The tradeoff between precision and recall is also well known in information retrieval and the "E" measure is one known measure that allows one to assign different weights to precision and recall and by doing so to give preference to one rather than the other. The "E" measure is defined as (see Ref. [18] for details) $((\beta^2 + 1)*\text{precision}*\text{recall})/(\beta^2\text{precision} + \text{recall})$. In our case, we tend to prefer precision to recall, as explained. To illustrate the actual meaning of such a preference, Figure 7 demonstrates the combination of precision and recall using the "E" measure when $\beta^2 = 0, 0.25, 0.5, 1, 2, 4$. $\beta^2 = 0.25$ means considering precision four times more important than recall. Looking at Fig. 7, it seems that, in general, the system is better at finding more visitors than identifying them accurately (performance improves as we prefer recall over precision). The practical meaning is that more effort should be made to improve the correct identification of the visitors.

It is worth noting that the experiment described above has certain limitations. It was conducted in a single, rather small, museum room, where the visit duration was relatively short. In this scenario the findings are quite limited. For example, there may be little benefit in adapting information to visitors after 40% of the visit. Moreover, we were interested in this stage to discover whether visitors exhibit consistent and predictable behavior, and hence we analyzed all visitors' logs in order to see how consistent their behavior is. However, museum visits, especially in large museums, may take quite a long time and therefore that it takes a few minutes to learn and identify the visitor is reasonable, and accurate identification can be achieved.

5 Conclusions and Future Work

This work analyzed the evolvement of museum visitors' behavior patterns during the visit. We first of all found an empirical validation of a qualitative model derived from ethnographic evidence. We then looked into the consistency of the behavior of the visitors and finally we investigated the possibility of predicting the type of visit from a few initial observations. The results showed that in certain cases it may be possible to identify the type of a museum visitor in the early stages of the visit. They also demonstrated the possible tradeoff between accurate categorization and comprehensive categorization. There are implications for the museum visit scenario. The knowledge of visitors' type may be used for tailoring the information delivered to them. Such an adaptation may make unnecessary questionnaire filling or interviewing at the beginning of a visit, which are common approaches for initial user model definition.

Future study should evaluate the practical meaning of the findings. Assuming visitors' type prediction is possible, as we suggest, then the practical meaning should be evaluated: how well the information is adapted to the visitors' needs based on such a prediction and whether they indeed appreciate this adaptation.

Additional aspects for future work are whether visitors' behavior stabilizes and then remains constant throughout the visit or whether it changes. If changes occur, then what may be the reasons for the change? How do different kinds of exhibits trigger a change in behavior? How does the length of the visit change the behavior (for instance, "*ANT*" visitors may get hungry and lose interest in the exhibits)?

Extending the scope even more, we need to consider small groups of visitors [11]. In many cases visitors visit the museum in small groups. How can inter-group type differences be identified and how can this information be used for improving the whole group's visit experience?

Furthermore, we need to examine how these technological solutions can be combined with recent museum research that motivates the need for supporting individual museum visitors, such as the identity-based model of Falk [7].

References

1. Baus J., Krüger A., and Wahlster W.: 2002. A resource-adaptive mobile navigation system. In Proceedings of the 7th international conference on Intelligent User Interfaces. San Francisco, CA.
2. Baus J., and Kray, C.: 2003. A Survey of Mobile Guides. Workshop on Mobile Guides at: Mobile Human Computer Interaction '03.
3. Boger Z., and Guterman, H.: 1997. Knowledge Extraction from Artificial Neural Networks Models. Proceedings of the IEEE International Conference on Systems Man and Cybernetics, SMC'97, Orlando, Florida, pp. 3030–3035.
4. Bohnert, F., Zukerman, I., Berkovsky, S., Baldwin, T., and Sonenberg, L.: 2008. Using interest and transition models to predict visitor locations in museums. *AI Commun.* 21, 2–3 (Apr. 2008), 195–202.

5. Cheverst K., Davies, N., Mitchell, K., Friday, A., and Efstratiou, C.: 2000. Developing a Context-aware Electronic Tourist Guide: Some Issues and Experiences. The CHI 2000 Conference on Human factors in Computing Systems, The Hague, Netherlands 17–24.
6. Chittaro L., and Ieronutti L.: 2004. A Visual Tool for Tracing Users' Behavior in Virtual Environments. Proceedings of the Working Conference on Advanced Visual Interfaces, Gallipoli, Italy 40–47.
7. Falk, H. J.: 2009. Identity and The museum visit experience. Walnut Creek, CA. Left Coast Press.
8. Hatala M., and Wakkary R.: 2005. Ontology-Based User Modeling in an Augmented Audio Reality System for Museums. User Modeling and User-Adapted Interaction. 15 pp. 339–380.
9. Kuflik, T., Callaway, C., Goren-Bar, D., Rocchi, C., Stock, O., and Zancanaro, M.: 2005. Non-Intrusive User Modeling for a Multimedia Museum Visitors Guide System. UM 2005, Edinburgh, UK. pp 236–240.
10. Kuflik T., and Rocchi. C.: 2007. User Modeling and Adaptation for a Museum Visitors' Guide – the PEACH Experience, in Stock & Zancanaro (eds.), PEACH – Intelligent Interfaces for Museum Visits, Springer-Verlag, Berlin-Heidelberg, pp 121–146.
11. Kuflik, T., Sheidin, J., Jbara, S., Goren-Bar, D., Soffer P., Stock O., and Massimo Zancanaro: 2007. Supporting Small Groups in the Museum by Context-Aware Communication Services. IUI 2007, Honolulu, Hawaii, USA, pp. 305–308.
12. Landis, J. R., and Koch, G. G.: 1977. The measurement of observer agreement for categorical data. Biometrics 33:159–174.
13. Marti, P., Rizzo, A., Petroni L., Tozzi, G., and Diligenti, M.: 1999. Adapting the Museum: A Non-intrusive User Modeling Approach. In: Proceedings of User Modeling Conference UM99.
14. Oppermann, R., and Specht, M.: 2000. A Context-Sensitive Nomadic Exhibition Guide. In proceedings of Handheld and Ubiquitous Computing: Second International Symposium, HUC 2000, Bristol, UK, pp. 127–142.
15. Petrelli, D., and Not, E.: 2005. User-Centred Design of Flexible Hypermedia for a Mobile Guide: Reflections on the HyperAudio Experience. User Modeling and User-Adapted Interaction: The Journal of Personalization Research 15(3–4). pp 303–338.
16. Sparacino, F.: 2002. The Museum Wearable: Real-Time Sensor-Driven Understanding of Visitors' Interests for Personalized Visually-Augmented Museum Experiences'. Museums and the Web, Boston, Massachusetts.
17. Stock, O., and Zancanaro, M.: 2007. PEACH: Intelligent Interfaces for Museum Visits. Cognitive Technologies Series, Springer, Berlin.
18. van Reijsbergen, C. J.: 1979. Information Retrieval. Butterworths.
19. Veron, E., and Levasseur, M.: 1983. Ethnographie de l'exposition, Paris, Bibliothèque Publique d'Information, Centre Georges Pompidou.
20. Zancanaro, M., Kuflik, T., Boger, Z., Goren-Bar, D., and Goldwasser, D.: 2007. Analyzing Museum Visitors' Behavior Patterns, In proceedings of the 11th International Conference on User Modeling, UM 2007 Corfu, Greece, pp. 238–246.
21. MacQueen, J.B.: 1967. Some Methods for classification and Analysis of Multivariate Observations, Proceedings of 5-th Berkeley Symposium on Mathematical Statistics and Probability, Berkeley, University of California Press 1:281–297.

Trust Management of Ubiquitous Multi-Display Environments

Ekatarina Kurdyukova, Elisabeth André, and Karin Leichtenstern

Abstract While a lot of research has been devoted to improving the security and the reliability of ubiquitous display environments, work on the user experience factor of trust is still scarce. To ensure that ubiquitous environments find acceptance among users, the user experience factor of trust should, however, not be underestimated. In this paper, we present a decision-theoretic approach to trust management that we consider particularly appropriate when a system has to balance the benefits and risks of a decision carefully. In the paper, we define decision policies that help maintain trust in critical situations, such as the loss of sensor data or the presence of unknown people. The approach has been employed in three interactive applications that have been developed as part of a university-wide ubiquitous displays management system.

1 Introduction

Recent years have produced a large variety of interactive displays that are installed in many public places. Apart from simply providing information (e.g., news or weather) to people in public places, such as coffee bars or airports, public displays make it possible for passing individuals to view, edit and exchange with each other specific data. Mobile phones represent a popular interaction device for interacting with these displays since they have been widely adopted by people as an everyday companion and can be customized to individual interaction preferences.

A social setting with the possibility of viewing personalized information in the presence of other people inevitably raises privacy issues. For example, a user may

E. Kurdyukova (✉) • E. André • K. Leichtenstern
Human-Centered Multimedia, Institute of Computer Science, Universitätsstr. 6a, Augsburg 86159, Germany
e-mail: Kurdyukova@informatik.uni-augsburg.de https://hcm-lab.de; Andre@informatik.uni-augsburg.de https://hcm-lab.de; Leichtenstern@informatik.uni-augsburg.de https://hcm-lab.de

A. Krüger and T. Kuflik (eds.), *Ubiquitous Display Environments*, Cognitive Technologies, DOI 10.1007/978-3-642-27663-7_11,
© Springer-Verlag Berlin Heidelberg 2012

be viewing personal information on a public display as other people pass by. The basic question that arises is how the system should react in such a situation. Should it trust the user will take the necessary actions herself? Or should it adapt autonomously to the changing social context, for example, by masking personal information on the public display? In the first case, there is the risk that the user expects the system to protect her privacy and is upset if no appropriate actions are taken. The limitation of the second approach is that it leads to an interruption of the user's work flow and might give her the feeling that she no longer has the system under control. Furthermore, there is the danger that the user does not understand the rationale behind the system's behavior and perceives the system as being only a little transparent.

The example illustrates that a system needs to balance the benefits and drawbacks of its actions carefully in order not to entail the risk that a user loses trust in its workings and does not use it any more.

In this paper, we present a decision-theoretic approach to a trust management system for ubiquitous display environments that assesses the user's trust in a system, monitors it over time, and applies appropriate measures to maintain trust in critical situations [12]. Such situations arise when, among other things, other people enter the user's private space [9], the system has to generate presentations based on inaccurate user or context data [4] or the system's adaptation behavior mismatches the user's expectations [3].

As a test bed for our research, we employ three applications that have been developed as part of a university-wide displays management system. Two applications run on public displays located in public rooms at Augsburg University. They can be operated and assisted by mobile phones. The first application, Friend Finder, is an interactive campus map that shows the current location and status of the user's friends. Since many students have difficulty orienting themselves on the campus (especially in new buildings), Friend Finder also supports a routing function, showing a detailed path to a selected friend. The second application, Media Wall, fosters exchange between the students. It represents a gallery of media items (pictures or videos) uploaded by students or scientific staff. Users can rank the media items, upload new items, and view their favorite ones. The third application created for a personal projector, On-Campus Navigator, is designed to enable indoor navigation within university buildings. Users can switch between two views on the mobile projector: an overview map with self-updating user position, and an arrow view that is projected in the user's physical environment and that points to the direction in which the user should move.

All three applications require sophisticated mechanisms to adapt to various trust-critical events. Since Friend Finder may disclose private information about a user's social network, it should be able to adapt intelligently to the surrounding social context in order to avoid possible threats to privacy. Ranking the media on Media Wall again may threaten the user's privacy in case of observation. Several users may interact with Friend Finder simultaneously, rendering their networks on the same campus map. Therefore, the system should be able to accommodate the data and interaction coming from multiple users. Systems such as On-Campus Navigator rely on sensor data to update the direction. Here, incomplete or possible incorrect sensor data may cause wrong or missing directions. The system should be able to cope with such deficits and adjust the presentation appropriately.

In the remaining chapter, we first present a scenario that served as inspiration for our applications. We then discuss work related to increasing the user's trust in ubiquitous display environments by appropriate interface design. After that, we present a model of a trust management system based on Bayesian Networks and influence diagrams and show how this model has been used within our applications. Finally, we present the first results of a user study.

2 Scenario

To illustrate our ideas, let us have a look at a scenario which inspired our applications.

Oscar is a student in the Informatics faculty. Currently he is having a lunch break and looking for his friends to join him for lunch at the University canteen. On a University floor, he finds a large public display. He has already used it once and has a mobile client installed on his PDA to operate the display. By means of one application, Oscar can find his friends on the campus map. Since Oscar is currently alone on the floor, he loads his personal social network on the public display. He notices that his best friends Barbara and Marina are both at the law library. Since he seldom visits the law faculty and does not know that campus area very well, he decides to download the route onto his mobile phone.

Suddenly a group of other students approach the display. Oscar is a bit irritated: he is not willing to expose his personal friends to the strangers, neither does he want to disclose his intention to meet specific persons. Apart from that, he considers the locations of his friends as private information not supposed to be shown to everybody. Surprisingly, he notices that the display masks the pictures of his friends, representing them by icons. The portraits and details have migrated now to his mobile device.

One of the approaching students wants to use the display as well. Thus, he stands next to Oscar and loads his personal social network on the campus map as well. The friends of the stranger are also rendered on the map; however, the icons are different. After Oscar selects Barbara on his mobile display, her icon is highlighted on the public screen. Oscar chooses the option to show the route, and the route is drawn on the large screen. Oscar then downloads the route to his mobile phone, and leaves the public display. His personal network disappears from the screen.

On his way, Oscar meets a friend, who is also willing to join him, Barbara and Marina for lunch. They both proceed to the law library, being navigated by the mobile application. For convenience, Oscar uses a personal projector attached to his mobile phone which facilitates on-campus navigation. The application displays an arrow pointing in the direction in which the user should move. Suddenly the arrow is disabled. The friends are irritated for a second. However, the application gives a short explanation that the signal is currently too low, and the update will take a couple of seconds. Indeed, in a moment, the arrow is revived, and the friends proceed to their destination.

3 Related Work

Most work that investigates trust issues in the context of ubiquitous displays environments focuses on the distribution of private and public data over various displays. Often mobile phones are used as private devices that protect the personal component of interaction from public observation.

Röcker et al. [9] conducted a user study to identify the privacy requirements of public display users. Based on the study, they developed a prototype system that automatically detects people entering the private space around a public display using infrared and RFID technology, and adapts the information that is visible based on the privacy preferences of the users. An evaluation of the system revealed that users are willing to use public displays when there is a mechanism for privacy protection.

Based on the evaluation of two mobile guides, Graham and Cheverst [4] analyzed several types of mismatch between the users' physical environment and information given on the screen and their influence on the formation of user trust. Examples of mismatches include situations where the system is not able to detect the user's current location correctly or situations where the system conveys a wrong impression about the accuracy of its descriptions. To help users form trust, Graham and Cheverst suggest employing different kinds of guide, such as a chaperone, a buddy or a captain, depending on characteristics of the situations, such as accuracy and transparency. For example, the metaphor of a buddy is supposed to be more effective in unstable situations than the chaperone or the captain.

Cao et al. [1] introduce the notion of crossmodal displays that enable users to access personalized information in public places while ensuring their anonymity. The basic idea is to display the main information publicly, but to add cues for individual users to direct them to information that is relevant to them.

All in all, there is a vivid research interest in the design of novel user interfaces for heterogeneous display environments. However, the few approaches that address the user experience factor of trust in such environments do not attempt to model the user experience of trust explicitly as a prerequisite for a trust management system. A number of approaches to modeling trust in computational systems have been presented. Especially in the area of multi-agent systems (MAS), trust models have been researched thoroughly (see, e.g., Castelfranci's and Falcone's introduction [2] to a formal modeling of trust theory and its applications in agent-based systems). However, these approaches either focus on trust in software components or aim at modeling trust in human behavior.

4 Dimensions of Trust

Much of the original research on trust comes from the humanities. Psychologists and sociologists have tried for a very long time to get a grasp on the inner workings of trust in interpersonal and interorganisational relationships. Other fields, such as economics and computer science, relied on their findings, but adapted them to the special requirements of their respective fields and the new context to which they are applied. There is consensus that trust depends on a variety of trust dimensions. However, there is no fixed set of such dimensions.

Trust dimensions that have been researched in the context of internet applications and e-commerce include reliability, dependability, honesty, truthfulness, security, competence, and timeliness (see, for example, the work by Grandison and Sloman [5] or Kini and Choobineh [6]). The more sociologically inclined authors [11] introduce willing vulnerability, benevolence, reliability, competence, honesty, and openness as the constituting facets of trust. Researchers working on adaptive user interfaces consider transparency as a major facet of trust (see, for example, the work by Glass et al. [3]).

Our set of trust dimensions is based on interviews with 20 computer science students who were asked to indicate the trust factors of user interfaces that they felt contributed to their assessment of trustworthiness. The most frequent mentions fell into the following categories: comfort of use ("should be easy to handle"), transparency ("I need to understand what is going on"), controllability ("want to use a program without automated updates"), privacy ("should not ask for private information"), reliability ("should run in a stable manner"), security ("should safely transfer data"), credibility ("recommendation of friends") and seriousness ("professional appearance").

A follow-up study revealed that there are statistically significant positive correlations between trust and the identified factors. The better the ratings for the trust dimensions, the better were also the ratings for trust. In addition, we observed a statistically significant positive correlation between the users' ratings of their general trust in software and their reported trust in the presented system. Furthermore, we found that a missing feeling of trust was accompanied by negative emotions, such as irritation, uneasiness, and insecurity. More information regarding this experiment and the exact findings can be found in Ref. [7].

5 Using a Decision-Theoretic Approach to Trust Management

In the following, we describe a model to assess the user's trust in a computer system. Our model of trust should account for the following characteristics of trust:

Trust as a subjective concept: There is a consensus that trust is highly subjective. A person who is generally confiding is also more likely to trust a software program. Furthermore, users respond individually to one and the same event. While some users might find it critical if a software asks for personal information, others might not care. We aim at a computational model that is able to represent the subjective nature of trust.

Trust as an uncertain concept: The connection between events and trust is inherently uncertain. For example, we cannot always be absolutely sure that the user notices a critical event at all. Furthermore, it may also happen that a user considers a critical event as rather harmless. As a consequence, it is not possible to predict with 100% certainty which level of trust a user has in a particular situation. A computational model of trust should be able to cope with trust as an uncertain concept.

Trust as a multifaceted concept: As shown in Sect. 4, trust is a multi-faceted concept. We therefore aim at a computational model that is able to represent explicitly the relative contribution of the trust dimensions to the assessment of trust. In particular, the model should help us predict the user's level of trust based on dimensions, such as the perceived transparency and controllability of a user interface. Furthermore, the model should allow us to add trust dimensions easily based on new experimental findings.

Trust as a dynamic concept: Trust depends on experience and is subject to change over time. Lumsden [8] distinguishes between immediate trust dimensions and interaction-based trust dimensions. Immediate trust dimensions, such as seriousness, come into effect as soon as a user gets in touch with a software system, while interaction-based trust dimensions, such as transparency of system behavior, influence the user's experience of trust during an interaction. To model trust as a dynamic concept, we need to be able to represent how the user's level of trust depends on earlier levels of trust.

Based on the considerations above, we have chosen to model users' feelings of trust by means of Dynamic Bayesian Networks. The structure of a Bayesian Network is a directed, acyclic graph (DAG) in which the nodes represent random variables while the links or arrows connecting nodes describe the direct influence in terms of conditional probabilities (see Ref. [10]).

5.1 Modeling the Determinants of Trust

Dynamic Bayesian Networks meet the requirements listed above very well. First of all, they allow us to cope with trust as a subjective concept. For example, we may represent the system's uncertain belief about the user's trust by a probability distribution over different levels of trust. Secondly, they enable us to model the non-deterministic nature of trust. In particular, we are able to make predictions based on conditional probabilities that model how likely it is that the child variable is given the value of the parent variables. For example, we may model how likely it is that the user has a moderate level of trust if the system's behavior is moderately transparent. Furthermore, Bayesian Networks enable us to model the relationship between trust and its dimensions in a rather intuitive manner. For example, it is

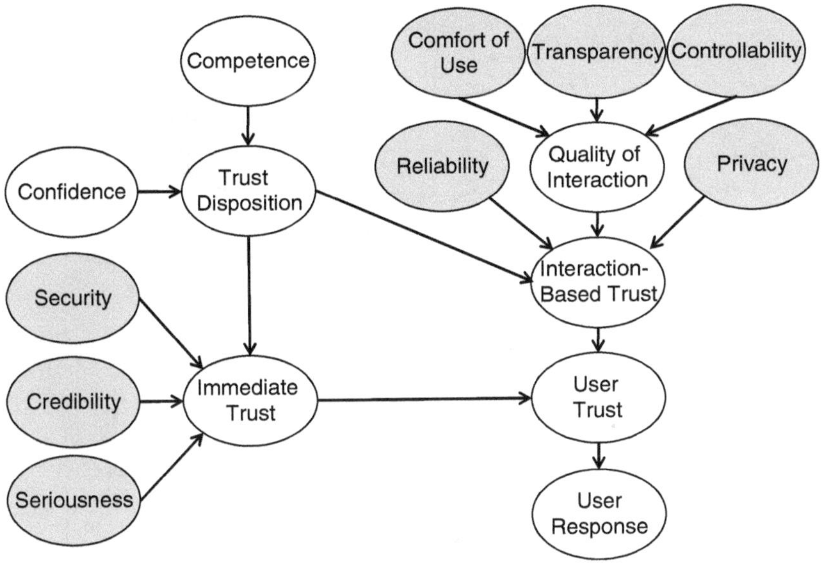

Fig. 1 Modeling trust by means of a Bayesian Network

rather straightforward to model that reduced transparency leads to a decrease in user trust. The exact probabilities are usually difficult to determine. However, the conditional probabilities can also be (partially) derived from the user data we collected in the experiment described in Ref. [7].

In Fig. 1, a Bayesian Network for modeling trust is shown. For each trust dimension, we introduced a specific node (highlighted in gray). Following Lumsden [8], we distinguish between immediate and interaction-based trust dimensions. Immediate trust dimensions include security (conveyed, for example, by the use of certificates), seriousness (reflected, for example, by the system's look-and-feel) and credibility (supported, for example, by company profile information). In this context, we would like to emphasize that trust dimensions may only affect the user's trust if the user is aware of them. For example, high security standards will only have an impact on user trust if the user knows that they exist. For the sake of simplicity, we assume that immediate trust dimensions do not change over time. That is we do not consider the fact that a user might notice references to security certificates only after working with a system over a longer period of time. To describe the determinants of interaction-based trust, we further distinguish between the quality of interaction, privacy and reliability. The quality of interaction is characterized by transparency, controllability and comfort of use. Both the establishment of immediate trust and interaction-based trust depend on the user's trust disposition which is characterized by his or her competence and their general confidence into technical systems.

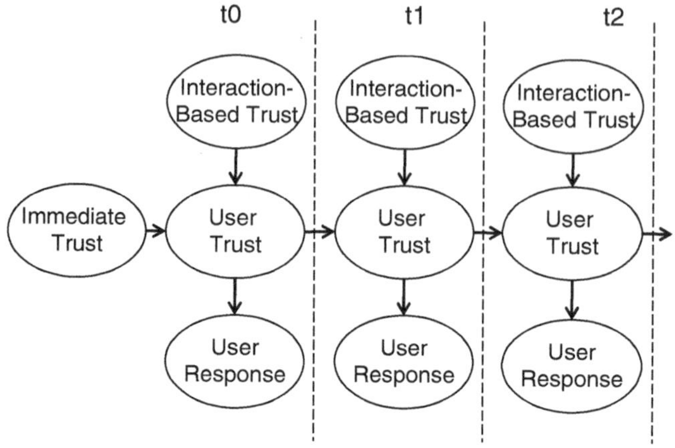

Fig. 2 Modeling the dynamics of trust by means of a Dynamic Bayesian Network

5.2 Monitoring Trust over Time

After smoothly interacting with a system over a longer period of time, the users' trust in a system is likely to increase. However, it may also happen that an unexpected system event, such as a sudden breakdown of the system, a substantial delay in the transfer of data or a serious leakage of data, causes a sudden loss of trust. All in all, the development of user trust must be continuously monitored at runtime in order to detect critical situations that require optimizations of the system to re-establish trust. As a consequence, we need not only a model that describes the relationship between user trust and its dimensions, but also one that explains the dynamics of trust.

Dynamic Bayesian Networks allow us to model the dependencies between the current states and the earlier states of variables. In particular, we are able to represent how the user's current level of trust is influenced by earlier levels of trust. In Fig. 2, a small portion of a Dynamic Bayesian Network is shown that illustrates how trust develops over time depending on the user's immediate level of trust and her interaction-based trust at time $t = 0$. Due to space limitations, we present only the time plates from ($t = 0$) to ($t = 2$). The arrow pointing from the node for user trust to the time plate for ($t = 2$) indicates that the user's trust at time ($t = 1$) influences the user's trust at time ($t = 2$). For simplicity, we consider only the user's level of trust at time t_{i-1} to determine the user's level of trust at time t_i.

5.3 Taking Decisions to Maximize Trust

So that we can use the Bayesian Network formalism for decision-making, it has to be extended to an influence diagram by adding a decision node and a utility node. The decision node represents all system actions that the system can perform, while the utility node encodes the utilities of all possible outcomes.

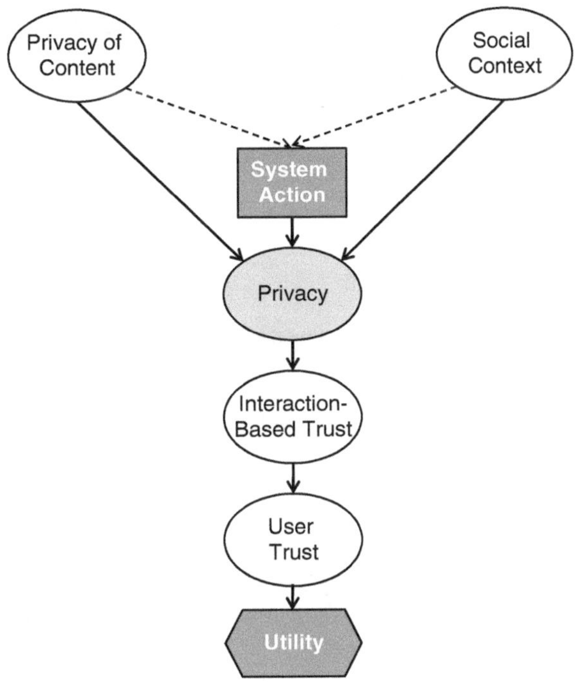

Fig. 3 Modeling decision making by means of influence diagrams

To make a decision, the system evaluates the utility of all possible options in terms of user trust and chooses the action with the highest utility. In Fig. 3, a small portion of the influence diagram is shown. We illustrate the basic idea by means of one trust dimension, namely privacy.

Privacy is handled as a hidden variable with three discrete values: low, medium and high. That is, its value cannot be directly observed, but has to be inferred from observable variables, such as *Privacy of Content* and *Social Context*. For example, the likelihood that the variable *Privacy* has the value *Low* would be high if *Privacy of Content* has the value *Private* and *Social Context* has the Value *People Approaching*. These dependencies are indicated by the arrows going from *Social Context* and *Privacy of Content* to *Privacy*. Associated with the decision node *System Action* is a table that describes the system's decision policy for each combination of the variables *Social Context* and *Privacy of Content*. The arrow going from *System Action* to *Privacy* represents the impact a particular system action, for example, the masking of private information, has on privacy.

6 Applying the Approach to Ubiquitous Displays

In the previous section, we described the general structure of an influence diagram as the basis for the implementation of a trust management system. In the following, we illustrate how to set the probabilities for a concrete application.

Table 1 Possible system reactions to trust-critical situations

Situation	Possible reaction
A – Changed social context	
User interacts with some private data on a public display. Another user approaches the display and starts to interact as well. Other people are around the display	A1: Do not take any action A2: Move all data from the public display to the mobile display A3: Mask private data on the public display and move the details to the mobile screen A4: Notify the user about potential privacy issues and offer options to protect data
B – Space conflicts	
Several users are approaching the public screen and want to interact with it. As a result, conflicts of space occur	B1: Allow just one user to interact at a time B2: Divide space among users B3: Move part of the data to mobile display, e.g., present final results on public displays and intermediate results on the mobile phone B4: One user interface integrates data of several users
C – Incomplete information	
The information the system needs to adapt a presentation to the user is not correct or incomplete	C1: Give textual explanations C2: Give visual feedback C3: Reduce level of detail C4: Give presentation based on available information

In an earlier experiment [7], we collected data for trust and its dimensions that we used for the creation of conditional tables employing the GeNIe (see http://genie.sis.pitt.edu) built-in algorithm for learning Bayesian Networks. To determine the parameters for the complete influence diagrams, we could, in principle, reply on empirical data as well. However, since the acquisition of these data turned out to be rather time consuming, we decided to set the probabilities that represent the dependencies between system actions and trust factors based on usability guidelines. For example, we know from the literature that there is a correlation between a system's transparency and user trust (see, for example, Ref. [3]).

In the following, we analyze the impact of various system reactions to typical trust-critical situations on relevant trust factors, providing illustrations from Friend Finder, Media Wall and On-Campus Navigator. Let us assume that the user is viewing private data on the public screen as other users pass by. Such a situation may occur in Friend Finder when users load a map of the university campus with friends on a public screen. The locations and pictures of friends are considered as private information, not supposed to be observed by just any one. Within the influence diagram shown in Fig. 3, this situation is described by the values of the variables *Social Context* and *Privacy of Content.*

In Table 1 (upper part), four possible responses to the described situation are listed. Basically, the system has to decide whether it should trust the user to take appropriate steps herself (Option A1), whether it should adapt the display of

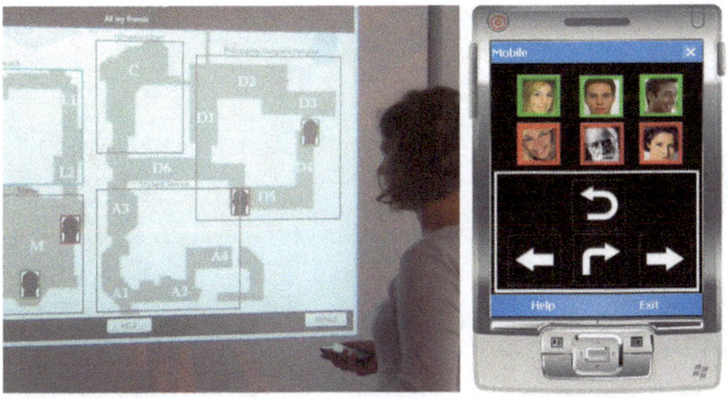

Fig. 4 Friend Finder: System masks private data on public display and displays the private details on the mobile screen

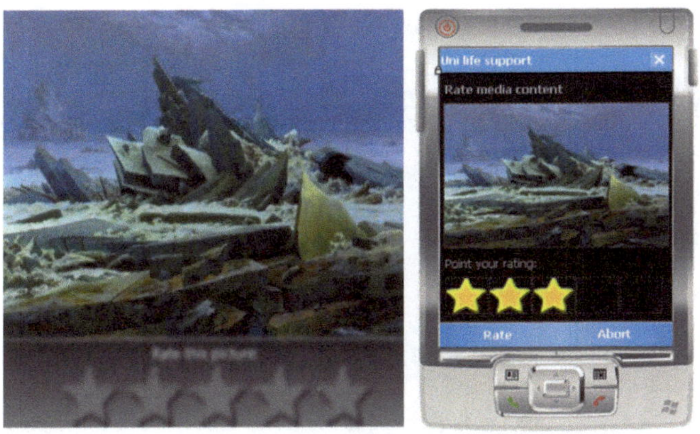

Fig. 5 Media Wall: The system masks user ranking and moves it to the mobile phone

information to the changed social context (Option A2 and A3) or whether it should offer the user options to protect private data (Option A4). Within the influence diagram shown in Fig. 3, the available options are represented by the decision node *System Action*.

Figure 4 shows Friend Finder's implementations of Option A3. The photos of the user's friends are masked with icons and the private data migrates to the mobile screen. Figure 5 shows the implementations of Option A3 in Media Wall. Here, the system masks the ranking and moves it to the mobile phone. Option A1 bears the risk that the user might expect the system to protect her privacy and will be upset if no appropriate actions are taken. Options A2 and A3 have the limitations that they cause an interruption of the user's work flow and might give her the feeling that she no longer has the system under control. The drawback of Option A4 is that there might not be enough time for the user to confirm the adaptations proposed by the system. Furthermore, it requires more effort from the user than the other options.

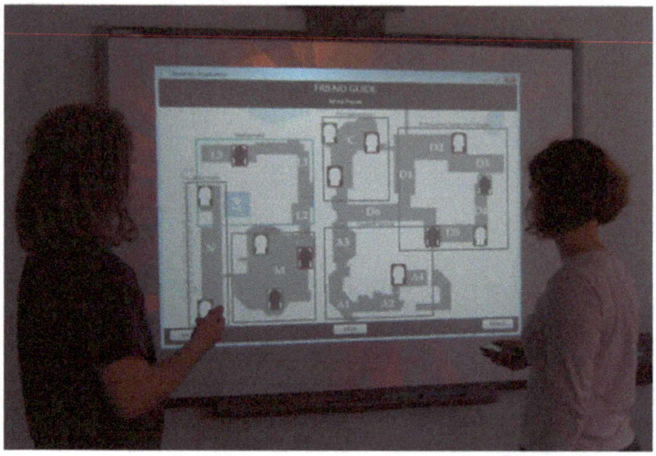

Fig. 6 Friend Finder: System masks the private data for both interacting users

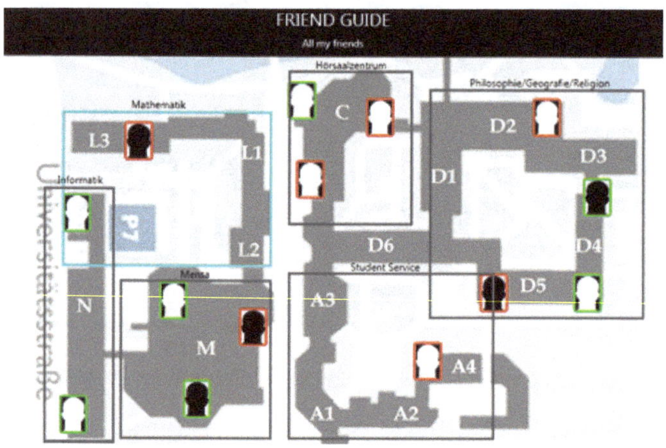

Fig. 7 Friend Finder: Sharing the public display by several users

Compared to Option A2, Option A3 has the advantage that the user still profits from the large real estate of the public screen, while preserving personal information. Furthermore, Option A3 allows several users to interact at the same time. In Fig. 6, two users browse their friends on a public display. Their network has been rendered on the same campus map. The display shows only the locations of the friends, and masks pictures with different icons for every user. The detailed information can be found on a mobile screen. The users distinguish their networks by the depiction of icons (Fig. 7).

When several people interact with public displays, conflicts of space are not uncommon. To solve such conflicts, the system may have just one user interact at the same time, allocate specific screen areas to the single users or move part of the

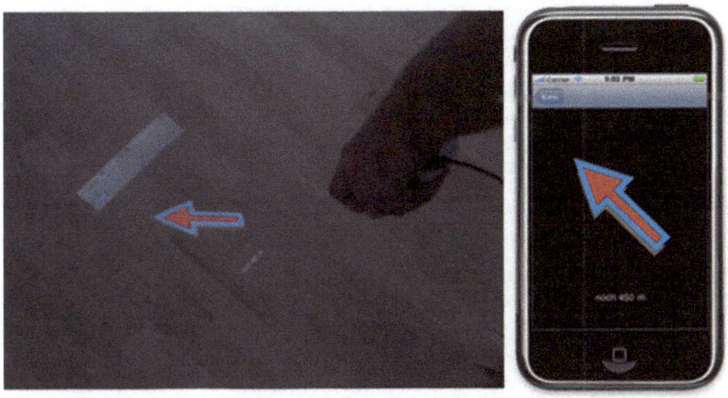

Fig. 8 On-Campus Navigator: Displaying a navigation arrow in the user's physical environment or on a mobile phone

data to mobile display (see the middle part of 1). Option B1 may affect the comfort of use negatively because the user may have to wait or postpone activities. On the other hand, the system's policy to handle conflicts of space may be easily understood. Option B2 may be advantageous when the display allows for a natural allocation of space that is immediately understood by users. In table top applications, a natural allocation of space is given by the user's position at the table. B3 requires the user to integrate public and personal information which may raise some usability problems. It can be advantageous when a common ground for user data can be found: for example, a map, or a table (see Fig. 6).

Finally, let us have a look at a situation where the system is forced to present data based on incomplete or incorrect information. The problems with incomplete or incorrect data often arise from sensor issues or unstable service functionality. In such a situation, the system has to decide whether it should inform the user about the problem or whether it should try to adapt its behavior to available data, for example, by reducing the level of detail (see lower part of Fig. 1). The drawback of the first option is that the user might lose trust in the system. The risk of the second approach is that the system might no longer be able to hide the problem at a later point in time resulting in an even greater loss of trust.

Figure 9 shows one of the strategies to cope with inaccurate data in On-Campus Navigator. As described earlier, On-Campus Navigator is able to give the user directions either by placing arrows in the user's physical environment using a miniaturized projector or by placing arrows on a mobile phone (Fig. 8). The On-Campus Navigator may temporarily lose sensor signal, or be unable to calculate an immediate direction. Once the system is not able to compute the direction, the arrow is temporarily disabled (Option C2) to indicate that the system does not have accurate location information. Another option to cope with the problem of insufficient signal is to provide an alternative view based on available information (Option C4). Thus, if the arrow direction is not possible to calculate, On-Campus Navigator may switch to the map view with the last available location update (Fig. 9).

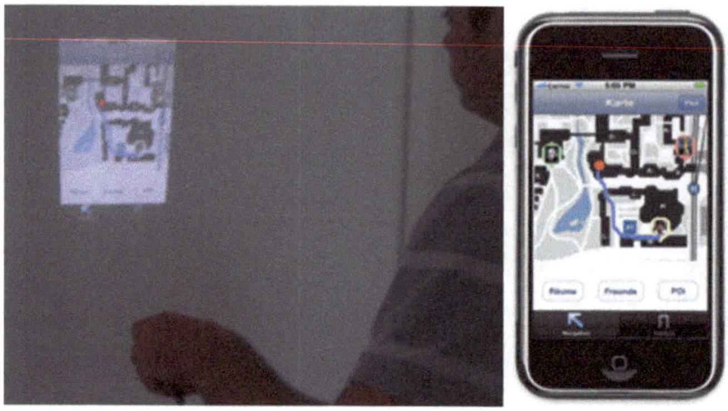

Fig. 9 On-Campus Navigator: Switching to the map view due to insufficient sensor data.

Table 2 Impact of possible system reaction on transparency (Tran), controllability (Contr), comfort of use (Comf), privacy (Priv) and reliability (Rel)

Action	Tran	Contr	Comf	Priv	Rel
Do not take any action (A1)	0	0	0	– –	– –
Complete migration to mobile phone (A2)	–	–	– –	++	0
Partial migration to mobile phone (A3)	–	–	–	+	0
Offer options to user (A4)	+	++	– –	++	0
Sequentialize interaction (B1)	0	0	–	0	0
Allocate space (B2)	0	0	–	0	0
Partial migration with use of common space (B3)	0	0	–	0	0
Textual or visual indication of error (B1, B2)	++	–	–	0	– –
Adapt presentation to error (C3, C4)	–	–	–	0	0

– – very likely to decrease, – likely to decrease, *0* not likely to change, + likely to increase, ++ very likely to increase

The decisions to be made by the system have an influence on the trust dimensions. The corresponding conditional tables were set up based on general guidelines in user interface design. For example, asking a user for confirmation before transferring private data to a public display has a positive impact on controllability, transparency and privacy, but a negative impact on ease of use. Further examples are shown in Table 2.

7 First Evaluation of the Approach

In the following, we present a first evaluation of the decision-theoretic approach.

First of all, we wanted to know to what extent the system is able to predict the user's ratings of trust based on her ratings of transparency, controllability, ease of use, seriousness, credibility, and security. To this end, we ask 20 users to rate five variants of a prototype that combined a public display (in this case a table top

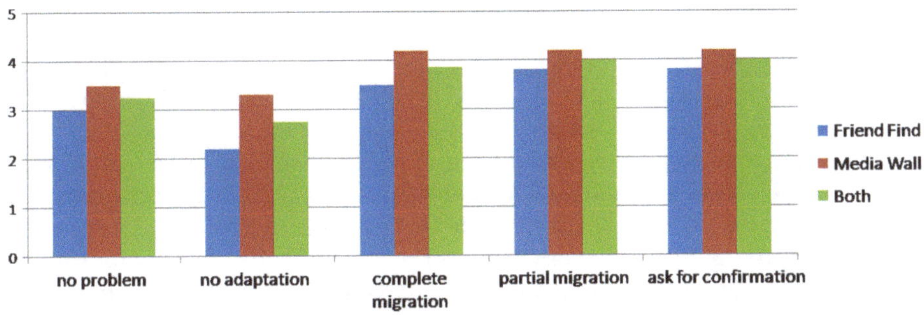

Fig. 10 Trust ratings for different responses

application) with a mobile phone. In particular, we produced a prototype that was less self-explainable (the interface included no help function and no descriptive labels), a second prototype that was less transparent (the system gave no reasons for its behavior), a third prototype that was less controllable (the system did not ask for user confirmations before executing an action), a fourth prototype that followed a less stricter privacy policy (the system displayed all kinds of data on user request on the table regardless of whether they were private or not) and finally a system that did not show any of these problems. The idea behind the variants was to obtain a sufficient variety of user ratings. To assess to what extent the user's level of trust can be predicted based on her ratings of transparency, controllability, ease of use, seriousness, credibility and security, we created a model using the Genie built-in algorithm for learning Bayesian Networks. When evaluating this model in tenfold cross validation, we achieved an accuracy rate of 73% for five classes (very low trust, low trust, medium trust, high trust, very high trust).

Secondly, we aimed at evaluating to what extent the decisions of the system contributed to a higher level of trust. As a first test, we presented six users in Friend Finder and Media Wall with a situation in which the user is viewing private information as other users pass by (Situation A). We furthermore showed them how the system would respond in such a situation (Options A1–A4) and asked them to rate their trust in the system on a scale from 1 (very untrustworthy) to 5 (very trustworthy). The results of this test are shown in Fig. 10. There was no clear preference for a particular system action. Despite the small number of users, the test shows, however, that the users clearly preferred the system to take the initiative and either adapt automatically to the social context or ask the user to confirm the adaptation. We take this as evidence that our adaptation policies were generally appropriate even though additional research is needed to prioritize decisions.

Another interesting finding is the fact that user trust was higher in situations where the system automatically adapted to a trust-critical situation than in situations where no problem occurred. Furthermore, the differences in the case of the Media Wall were less extreme than in that of the Friend Finder. We hypothesize that the information displayed with Media Wall (ratings of photos) was considered as less sensitive than the information displayed with Friend Finder (location of people).

8 Conclusions

Ubiquitous displays environments require a high degree of flexibility due to the changing social context and the probably incomplete or inaccurate information on which a system has to base its presentations. In order to maintain user trust in such environments, a system needs to be able to evaluate carefully the consequences of its actions and the trade-offs between them. In this paper, we presented a decision-theoretic approach to trust management. The approach has been informed by guidelines on user interface design in order to assess the impact of system actions on trust dimensions, such as comfort or use, transparency, controllability and privacy. A first evaluation of the approach within two applications that have been developed as part of a university-wide public displays environment revealed that users preferred the adaptive to the non-adaptive system.

Our future work will concentrate on conducting further experiments considering a larger number of users and a greater variety of trust-critical situations. The experiments conducted so far seem to indicate that the decisions taken by the system to adapt to a trust-critical situation find general user acceptance. However, a greater amount of user data is necessary to evaluate the appropriateness of particular design decisions. So far, the presented decision-theoretic approach and the corresponding evaluation concentrated on short-term interactions with public displays. We need, however, to take into account that users interacting with a system for a longer period of time might become annoyed about repeated system explanations and requests to confirm system actions. Thus, a too high amount of controllability and transparency may even negatively affect user trust. In order to avoid such problems, our future work will also consider the history of interactions during trust management.

Acknowledgement This research is partly sponsored by *OC-Trust* (FOR 1085) of the German research foundation (DFG).

References

1. H. Cao, P. Olivier, and D. Jackson. Enhancing privacy in public spaces through crossmodal displays. *Soc. Sci. Comput. Rev.*, 26(1):87–102, 2008.
2. C. Castelfranchi and R. Falcone. *Trust Theory: A Socio-Cognitive and Computational Model.* Wiley, 2010.
3. A. Glass, D. L. McGuinness, and M. Wolverton. Toward establishing trust in adaptive agents. In *IUI '08: Proceedings of the 13th international conference on Intelligent user interfaces*, pages 227–236, New York, NY, USA, 2008. ACM.
4. C. Graham and K. Cheverst. Guides, locals, chaperones, buddies and captains: managing trust through interaction paradigms. In *3rd Workshop 'HCI on Mobile Guides' at the Sixth International Symposium on Human Computer Interaction with Mobile Devices and Services*, pages 227–236, New York, NY, USA, 2004. ACM.
5. T. Grandison and M. Sloman. A survey of trust in internet applications. *IEEE Communications Surveys and Tutorials*, 3(4):2–16, 2000.

6. A. Kini and J. Choobineh. Trust in electronic commerce: definition and theoretical considerations. In *Proc. of the Hawaii International Conference on System Sciences*, volume 31, pages 51–61, 1998.
7. K. Leichtenstern, E. André, and E. Kurdyukova. Managing user trust for self-adaptive ubiquitous computing systems. In *Proceedings of the 8th International Conference on Advances in Mobile Computing and Multimedia*, MoMM '10, pages 409–414, New York, NY, USA, 2010. ACM.
8. J. Lumsden. Triggering trust: to what extent does the question influence the answer when evaluating the perceived importance of trust triggers? In *BCS HCI '09: Proceedings of the 2009 British Computer Society Conference on Human-Computer Interaction*, pages 214–223, Swinton, UK, UK, 2009. British Computer Society.
9. C. Röcker, S. Hinske, and C. Magerkurth. Intelligent privacy support for large public displays. In *Proceedings of Human-Computer Interaction International 2007 (HCII'07)*, Beijing, China, 2007.
10. S. J. Russell and P. Norvig. *Artificial Intelligence a modern approach.* Prentice Hall, Upper Saddle River, N.J., 2nd international edition, 2003.
11. M. Tschannen-Moran and W. Hoy. A multidisciplinary analysis of the nature, meaning, and measurement of trust. *Review of Educational Research*, 70(4):547, 2000.
12. Z. Yan and S. Holtmanns. Trust modeling and management: from social trust to digital trust. *Book chapter of Computer Security, Privacy and Politics: Current Issues, Challenges and Solutions*, 2008.

From Research to Practice: Automated Negotiations with People

Raz Lin and Sarit Kraus

Abstract The development of proficient automated agents has flourished in recent years, yet making the agents interact with people has still received little attention. This is mainly due to the unpredictable nature of people and their negotiation behavior, though complexity and costs attached to experimentation with people, starting from the design and ending with the evaluation process, is also a factor. Even so, succeeding in designing proficient automated agents remains an important objective. In recent years, we have invested much effort in facilitating the design and evaluation of automated agents interacting with people, making it more accessible to researchers. We have created two distinct environments for bargaining agents, as well as proposing a novel approach for evaluating agents. These are key factors for making automated agents become a reality rather than remain theoretical.

1 Introduction

Negotiation is the process in which interested parties exchange information in order to come to an agreement [3]. Thus, negotiation surrounds our every day life. This emphasizes the need for proficient automated negotiation agents which can interact with people in open environments. The introduction of open environments

This research is based upon work supported in part by the U.S. Army Research Laboratory and the U.S. Army Research Office under grant number W911NF-08-1-0144 and under NSF grant 0705587.

R. Lin (✉)
Department of Computer Science, Bar-Ilan University, Ramat-Gan 52900, Israel
e-mail: linraz@cs.biu.ac.il

S. Kraus
Department of Computer Science, Bar-Ilan University, Ramat-Gan 52900, Israel

Institute for Advanced Computer Studies, University of Maryland, College Park, MD 20742, USA
e-mail: sarit@cs.biu.ac.il

A. Krüger and T. Kuflik (eds.), *Ubiquitous Display Environments*, Cognitive Technologies, DOI 10.1007/978-3-642-27663-7_12,
© Springer-Verlag Berlin Heidelberg 2012

is important, as the automated agent needs to be able to interact with different types of people who have different characteristics, originate from different countries and cultures and so on. Automated negotiation agents capable of negotiating proficiently with people must therefore deal with the fact that people are diverse in their behavior and each individual might negotiate in a different manner.

Two main difficulties in the design of proficient automated agents that interact with people stem from the fact that people are involved. First, the design is not straightforward. It is not sufficient to design agents that follow economical paradigms and strive for the equilibrium, or to design agents that implement quantitative decision-making approaches and are expected utility maximizers, as results from social sciences suggest that people do not follow equilibrium strategies [4, 14]. Human behavior is diverse and cannot be captured by a monolithic model. Humans tend to make mistakes, and they are affected by cognitive, social and cultural factors [1, 8]. Moreover, when playing with people, the theoretical equilibrium strategy is not necessarily the optimal strategy [20]. In this respect, equilibrium-based automated agents that play with people must incorporate heuristics to allow for "unknown" deviations in the behavior of the other party. In addition to the problems of the design of a proficient agent, the researcher also needs to evaluate the agent's interactions with people. The scientific evaluation of computational strategies for human-computer decision-making is a costly process, involving time, effort and personnel.

In recent years, we have focused our efforts in making this process practical and reachable. In this paper, we will describe two environments we have designed, the GENIUS environment [11], which is a bilateral negotiation environment, and the Colored Trails game environment [7], which is a game played on a *nxm* board of colored squares. These two environments allow the test-bedding of automated agents, as well as the matching of agents with other agents, people with people and people with automated agents. However, as we mentioned, designing agents does not suffice. Thus, we also showed how the evaluation process of the automated agents can be facilitated using peer-designed agents. Building on these, we hope to motivate more researchers in pursuing agents design and evaluation with people.

The remainder of this paper is organized as follows. In Sect. 2, we provide an overview of the GENIUS environment and describe an automated negotiator designed in GENIUS. We continue and describe the Colored-Trail environment and an automated agent designed in this environment in Sect. 3. In Sect. 4, we provide our novel approach for facilitating the evaluation process of automated agents. Finally, we provide a summary and discuss the results.

2 The GENIUS Environment

GENIUS [11] is a *G*eneral *E*nvironment for *N*egotiation with *I*ntelligent multi-purpose *U*sage *S*imulation. With the constant introduction of new domains, e-commerce and other applications which require automated negotiations, generic automated negotiators encompass many benefits and advantages over agents that are designed for a specific domain. GENIUS both assists in the *design* of strategies for automated negotiators and also *supports* the evaluation process of the agent. Thus, we believe this tool is very useful for agent designers and can take a central part in the process of designing automated agents. While designing agents can be done in any agent-oriented software engineering methodology, GENIUS wraps this in an easy-to-use environment and allows the designers to focus on the development of *strategies* for negotiation in an open environment with multi-attribute utility functions.

GENIUS incorporates several mechanisms that aim to support the design of a general automated negotiator. The first mechanism is an analytical toolbox, which provides a variety of tools to analyze the performance of agents, the outcome of the negotiation and its dynamics. The second mechanism is a repository of domains and utility functions. As agents are built using the same infrastructure of GENIUS, GENIUS maintains a repository of automated agents and enables the evaluation and comparison of different strategies used by the automated agents. This is an important contribution as it allows researchers to empirically and *objectively* compare their agents with others in different domains and settings. This is an important contribution with respect to the validation of results reported by researchers with regard to their automated negotiators.

GENIUS supports a multi-issue negotiation setting in which participants engage in repeated negotiation rounds in which they can propose an agreement which consists of values for a subset or all of the issues in the negotiation. The negotiation protocol proceeds as follows: At each time period each participant can propose a possible agreement for some subset of the issues (see Fig. 1), and the other participant can either accept the offer, reject it, or opt out. The protocol is an extension of the classic alternating offers protocol of Osborne and Rubinstein ([16], pp. 118–121) to support partial agreements as well as an opting-out option.

The negotiation terminates when one of the following holds: (a) the negotiators reach an agreement for all of the issues, (b) one of the participants opts out, thus forcing the termination of the negotiation with an default outcome, or (c) a predefined deadline (dl) is reached, whereby, if a partial agreement is reached it is implemented or, if no agreement is reached, a status quo (SQ) outcome is implemented. The utility for participants depends on the agreement and the time period in which it was reached.

Formally, let I denote the set of issues in the negotiation, O_i the finite set of values for each $i \in I$ and O a finite set of values for all issues ($O_1 \times O_2 \times \ldots \times O_{|I|}$). We allow partial agreements, $\perp \in O_i$, for each $i \in I$. Therefore an offer is denoted as a vector $\vec{o} \in O$.

The negotiation session is divided into time periods, $\textbf{Time} = \{0, 1, \ldots, dl\}$. Each agent is assigned a time cost which influences its utility as time passes. In each period $t \in \textbf{Time}$ of the negotiation, if the negotiation has not terminated earlier, each agent can propose a possible agreement and the other agent can either accept the offer, reject it or opt out.

The negotiation problem also involves incomplete information about the preferences of the opponent. We assume that there is a finite set of agent types.

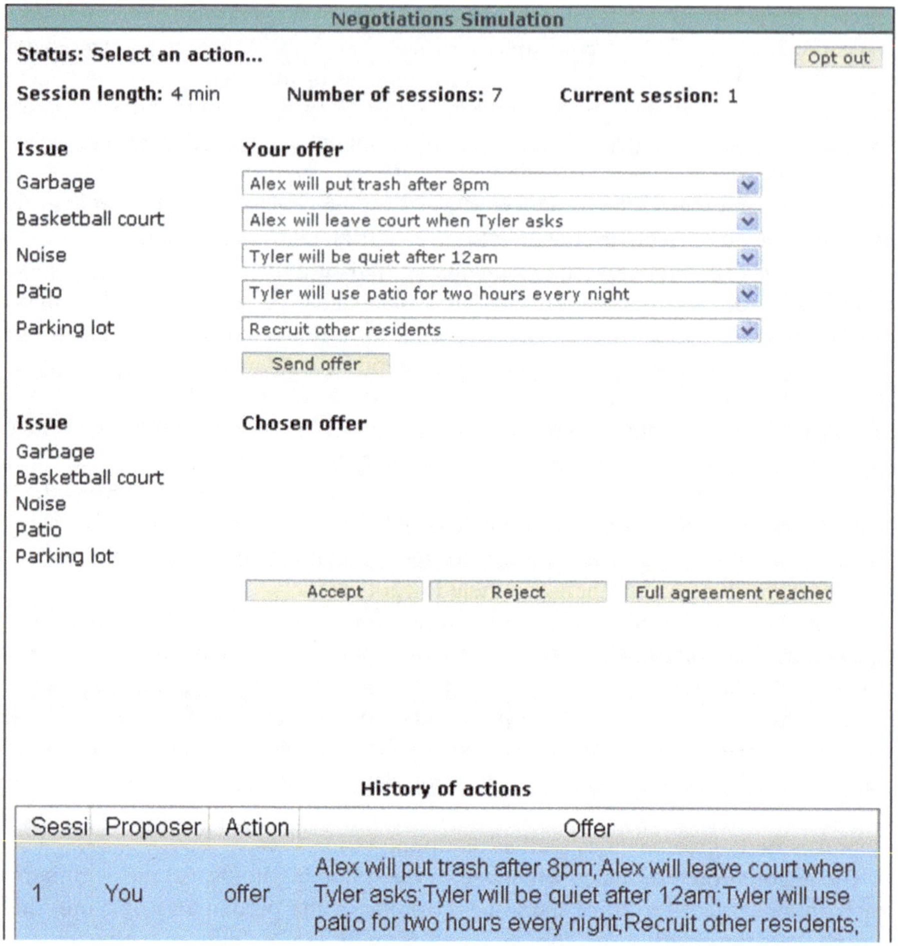

Fig. 1 Bilateral negotiation: generating offers screen

These types are associated with different additive utility functions (e.g., one type might have a long term orientation regarding the final agreement, while the other type might have a more constrained orientation). Formally, we denote the possible types of agents as $\mathbf{Types} = \{1, \ldots, k\}$. Given $l \in \mathbf{Types}$, $1 \leq l \leq k$, we refer to the utility of an agent of type l as u_l, and $u_l : \{(O \cup \{SQ\} \cup \{OPT\}) \times \mathbf{Time}\} \rightarrow \mathbf{R}$. Each agent is given its exact utility function. The negotiators are aware of the set of possible types of the opponent. However, the exact utility function of the rival is private information.

In the next subsection, we describe the *KBAgent* that was designed and evaluated using this system in two scenarios: an employer and an employee negotiating over terms of a job contract and a diplomatic negotiation process between two countries. Detailed score functions for both domains can be found in Ref. [12].

2.1 *The* KBAgent

The *KBAgent* [17] is an automated negotiator that was proven to be proficient when negotiating with people. It incorporates machine learning techniques to allow it to generate a general model of the opponent and also follows qualitative decision-making, rather than quantitative. The *KBAgent* negotiates with each person only once, and uses a database of past negotiation sessions between specific agents (types of negotiators) to allow it to be more proficient in negotiations with agents of that specific type (in our experiments, we used a database of past negotiations of people to allow it to be matched with people). Based on a database with past negotiation sessions, the agent performs offline learning, which is based on the kernel-based density estimation ([21], Chap. 2). *KDE* is a technique for estimating the probability functions based on samples, and can also be viewed as a smoothing technique based on samples. An advantage of this technique is that it does not assume a model of the other side, just like the data we expect to have in our database. From the database, the agent estimates the probability of an offer to be accepted, the probability of it to be offered and the expected average utility for the other party. These probabilities are then used in its decision-making component, either when accepting an offer or to determine the agent's concession rate. Using the kernel-based density estimation the *KBAgent* is capable of using even small databases and does not have to rely on many past negotiation sessions. The *KBAgent* has been shown to negotiate proficiently with people [17]. In the remainder of this section we elaborate on the design of the *KBAgent*.

General opponent modeling component: This component is responsible for generating the general opponent modeling. The *KBAgent* uses the *KBA* method to estimate the probability of each offer to be proposed by the other party during a *given turn*. In addition, the database is used to calculate the probability of each offer to be accepted by the other side during any turn. We denote by $Q(\vec{o})$ the estimated probability of each offer $\vec{o} \in O$ to be accepted at any given time by the opposite party and by $P(\vec{o}, t)$ the probability that the opposite party would propose offer $\vec{o}$ during a given turn t. These probabilities will be used later in the decision-making process.

The database of past negotiations includes logs of past negotiation sessions between two sides in the specified domain. The negotiation sessions include all offers made by the two parties in each turn and whether or not the offers were accepted. The negotiation sessions may be collected from any population, not necessarily similar to the one with which the *KBAgent* negotiates.

The estimation of the probabilities is done separately for each possible agent type. If the negotiation sessions in the database are not labeled with the type of the agent, the type is elicited using a simple Bayes' classifier (similar to the one used for estimating the believed type of the other party during the negotiations). To estimate the probability of offers being proposed by the agent of a given type, we extract from the database all offers proposed by the agents of that type in any given turn t, $1 \leq t \leq dl$. To apply the *KDE* algorithm all offers have to be assigned a unique numerical value, thus we order the offers by the utility values of the other side and rank them, such that the offer with the highest possible utility value is ranked 1. Then the *KDE* is applied and the results also include a smooth probability value even for agreements that were not part of the samples in the database.

To estimate the probability that the other side will accept an offer during the negotiation we extract from the database all the offers ever accepted or proposed by at least one negotiator playing the role of the other party (under the assumption that if anyone had proposed an offer they would also accept it when offered to them). We will refer to them as "acceptable offers". Note that we actually use the *KDE* method to obtain estimation on the utility of offers, not to learn any estimation on the offers themselves. For each offer during the negotiation that the *KBAgent* proposes, the acceptance probability is calculated by computing the ratio between (a) the number of the offers from the acceptable offer list that have a lower utility value for the other party than the utility value of the proposed offer, and (b) the total number of offers in the acceptable list of the other party.

We use the following example to demonstrate this. Assume that the database contains sessions in which people proposed offers with the following utilities: {400, 380, 300, and 200}. Also assume that they accepted an offer with a utility value of 280. Thus, the acceptable offers list will contain the offers with the following utility values: {400, 380, 300, 200, and 280}. Now assume that we would like to calculate the probability that the other party will accept an offer with a utility value of 290. Since there are two agreements in the acceptable list with a utility value lower than 290 (280 and 200) the estimated probability of acceptance is $2/5 = 0.4$.

Finally, from the database we calculate the average expected utility of the other party. This is done by averaging the final utility scores of all agreements reached in the negotiation sessions (regardless of during which turn the agreements were reached). We denote this average as *ExpectedOppAvg*.

The decision-making component: The decision-making valuation component takes into account the agent's utility function, as well as the believed type of the opponent (note that the believed type of the opponent is also influenced by the offers proposed by the opponent). This data is used both for deciding whether to accept or reject an offer and for generating an offer. In our settings, although several offers can be proposed each time period, we restrict the agent to a single offer in each period. This is done due to the fact that our mechanism for generating offers only produces one distinct offer in a given time period. The opponent, on the other hand, is free to propose several offers, and the agent can respond to all the offers, which actually occurred in the experiments.

Generating offers: The *KBAgent* implements a concession-oriented strategy when generating offers. The process of deciding what offers to propose during each turn consists of three phases. First, a list of offers ranked by their *QOValue* is generated. The *QOValue* of an offer is an alternative to the Nash bargaining solution. In short, it tries, in a qualitative manner, to evaluate the offers based on the agent's utility and based on the likelihood of their acceptance by the other party (c.f. Ref. [12] for more details). The *QOValue* of an offer is calculated using the following formula:

$$\begin{aligned} QOValue(\vec{o}) &= \min\{\alpha_o, \beta_o\} \\ \text{where } \alpha_o &= rank_a(\vec{o}) \cdot lu_a(\vec{o}) \\ \text{and } \beta_o &= [lu_a(\vec{o}) + lu_b(\vec{o})] \cdot rank_b(\vec{o}) \end{aligned} \tag{1}$$

where *rank*(·) is the ranking value of an offer, which is associated with each offer and a given utility function u. The rank number of an offer $\vec{o} \in O$ is calculated using the following formula:

$$rank(\vec{o}) = \frac{order(\vec{o}, O)}{|O|} \tag{2}$$

where *order*(·,·) is the ordering function which places the offer $\vec{o}$ in an ordinal scale between 1 and $|O|$ according to its utility value compared to all other offers in O. *lu*(·) denotes the Luce number of an offer [13], which is a non-negative number that is associated with each offer. The Luce number of an offer $\vec{o} \in O$ is calculated using the following formula:

$$lu(\vec{o}) = \frac{u(\vec{o})}{\sum_{\vec{x} \in O} u(\vec{x})} \tag{3}$$

After calculating the *QOValue* of each offer, the *KBAgent* orders all offers by their *QOValue*. The offer with the maximal *QOValue* is the first offer the *KBAgent* proposes. The second phase is to construct a new list which will be the one from which offers will be proposed. This new list is based on the sorted offers by their *QOValue*, in which the utility values for the *KBAgent* are above the value of the status quo. In addition, the *KBAgent* discards all offers with a lower *QOValue* that do not improve the other party's utility from offers with a higher *QOValue*. That is, if $QOValue(\vec{o}_i) < QOValue(\vec{o}_j)$ and $u_{opp}(\vec{o}_i) < u_{opp}(\vec{o}_j)$ the offer $\vec{o}_i$ will be removed from the proposal list. The last phase is deciding which offer to propose each turn. As mentioned earlier, the first offer that is proposed is the one with the maximal *QOValue*. The other offers are picked from the ordered list based on the concession rate applied by the *KBAgent* and are chosen with a decreasing *QOValue* for the agent and an increasing utility value for the other party. Note that while we assume that the other party is not rational, we refer to a case where it does not maximize its expected utility, thus it should still prefer to accept offers with higher utility values than lower ones.

The concession rate is determined such that after passing 80% of the negotiation turns, the *KBAgent* will propose the offer closest to the *ExpectedOppAvg*, that is, the average expected utility value of the other party, as estimated from the database of

Table 1 Example of deciding which offers will be proposed

Offer Idx	$QOValue(\vec{o})$	$u_{opp}(\vec{o})$	Ordered list of offers
0	13.45	**350**	**0**
1	12.5	300	–
2	12	**400**	**1**
3	11.22	**430**	**2**
4	10.3	350	–
5	10	**435**	**3**
6	9.87	470	4
7	9.8	**490**	**5**
8	9	410	–
9	8.8	500	6

Proposed offers are marked in bold

past negotiations. In this way the *KBAgent* makes concessions while offering agreements which are still efficient for it. The concession rate basically determines the order in which offers in the offer list will be proposed. Thus, even if after 80% of the turns the other side rejects the offer, the *KBAgent* continues to propose offers from its list based on the concession rate.

Listing 1 describes the pseudo-code of the algorithm for generating the proposed offers and the concession rate.

We will demonstrate *KBAgent*'s offer generation mechanism using the following example. Assume that in a given negotiation session there are five turns and ten possible offers. Table 1 lists the possible offers sorted by their *QOValue*. Based on the utility value of each offer for the other party, it also lists the final ordered list of offers from which the *KBAgent* will propose offers to its counterpart. Now, assume that the average expected utility value of the opposite side, as estimated from the database, is 440. Based on *KBAgent*'s algorithm, after 80% of the turns, that is, after turn 4 ($5 \cdot 0.8 = 4$), it should propose offer #7, which is the first one with a higher utility value for the other party than the average expected utility determined from the database. This offer is the fifth offer in the ordered list of offers and thus the agent can now establish its concession rates and the offers to be proposed each turn. Since the *KBAgent* aims to propose the fifth offer after turn 4, the concession rate is determined as $5/4 = 1.25$. Thus, the agent should first propose the first offer in the list and then continue using the concession rate. That is, it should propose offers from its list in the following order: 0, 1.25, 2.5, 3.75, 5, 6.25. The numbers are rounded down, so the *KBAgent* eventually proposes the offers ordered 0, 1, 2, 3, 5 in its offer list, in that order. These offers are also marked in bold in Table 1.

Accepting offers: An important aspect of the agent's strategy is the decision of whether to accept or reject an offer. The *KBAgent* determines a time dependent threshold to decide whether to accept or reject an offer. Obviously, a good threshold should be used and not an arbitrary one. In order to decide on the optimal threshold, the probabilities learned from the database of past negotiations are used.

Listing 1 Generating Proposed Offers

```
for all o⃗ ∈ Database do
  qoValue = QOValue(o⃗)
  Insert o⃗ to QOSortedList
end for
QuickSort (QOSortedList) based on qoValue
Insert QOSortedList(0) to OfferList
j = 0, maxUtil = u_opp (QOSortedList(0)), concessionIndex = 0
for i =1 to QOSortedList.length-1 do
  util = u_opp(o⃗_i)
  if (util>maxUtil) and (u_KB(o⃗)>u_KB(SQ)) then
    Insert QOSortedList (i + 1) to OfferList
    maxUtil = util
    j = j + 1
    if (concessionIndex = 0) and (util > ExpectedOppAvg) then
      concessionIndex = j
    end if
  end if
end for
concessionRate = concessionIndex/(0.8 · |Time|)
```

Let $o_{KB}(t)$ be the offer proposed by the *KBAgent* at time t. The acceptance threshold for the *KBAgent* is calculated per every turn t of the negotiation and denoted by α_t. We want α_t to allow for maximization of the expected utility value of the agent for turn t. Thus, we first calculate the expected utility value of the *KBAgent* for each turn of the negotiation, denoted $E(t, \alpha_t)$. Since the negotiation terminated at turn dl we can use backward induction to calculate this. The expected utility of the agent if the deadline is reached and no agreement has been made equals the utility of the status quo. Thus:

$$E(dl, \alpha_{dl}) = u_{kb}(SQ) \tag{4}$$

In the preceding turn, if *KBAgent*'s offer was not accepted, it should accept any agreement with a utility value higher than *SQ*. Otherwise, the negotiation will terminate with a status quo outcome and a lower utility value for the agent. Thus, its expected utility depends on the probability that offers with a utility value higher than the status quo would be proposed by the other party. For all offers which are above the acceptance threshold of the *KBAgent*, we sum their probability of being proposed by the counterpart multiplied by their utility value. For all other agreements, as the agent will reject them, we sum the probability of the counterpart proposing them multiplied by the value of the status quo. Formally,

$$\begin{aligned} &E(dl-1,\ \alpha_{dl-1}) \\ &= \sum_{u_{KB}(\vec{o},dl-1) \geqslant \alpha_{dl-1}} P(\vec{o},\ dl-1) u_{KB}(\vec{o},\ dl-1) \\ &+ \sum_{u_{KB}(\vec{o},dl-1) < \alpha_{dl-1}} P(\vec{o},\ dl-1) u_{KB}(SQ,\ dl-1) \end{aligned} \tag{5}$$

For every other turn, we calculate the expected utility of the *KBAgent* using a recursion. Basically, the expected utility of the agent at turn *i* depends on whether or not it accepts the proposed offer:

- For all offers above the agent's acceptance threshold, the *KBAgent* sums the multiplication of their utility value with the probability that they will be proposed by its counterpart.
- For all offers below the agent's acceptance rate, the *KBAgent* calculates the probability that its offer will be accepted by its counterpart in the next turn or, if it is rejected, the agent's expected utility for the subsequent turn.

Formally,

$$\begin{aligned} E(t, \alpha_t) = & \sum_{u_{KB}(\vec{o}, t) \geqslant \alpha_t} P(\vec{o}, t) u_{KB}(\vec{o}, t) \\ & + \left(\sum_{u_{KB}(\vec{o}, t) < \alpha_t} P(\vec{o}, t) \right) \{ Q\left(o_{KB}(t+1)\right) \cdot u_{KB}(o_{KB}(t+1), t+1) \\ & + [1 - Q(o_{KB}(t+1))] E(t+1, \alpha_{t+1}) \} \end{aligned} \tag{6}$$

The last calculation needed is the extraction of α_t to determine the acceptance threshold for each turn. To this end, we use the derivative of Formula 6 by α_t,[1] and as we would like to maximize the expected utility we find the α_t value when the derivative equals 0. Thus, we need to find the solution for the following equation:

$$\begin{aligned} \alpha_t = & Q(o_{KB}(t+1)) u_{KB}(o_{KB}(t+1), t+1) \\ & + (1 - Q(o_{KB}(t+1))) E(t+1, \alpha_{t+1}) \end{aligned} \tag{7}$$

Based on Formulas 5, 6 and 7 we can backtrack and calculate the optimal acceptance threshold for each turn. Note that these calculations are done separately for each different believed agent type of the counterpart.

Identifying the opponent's type: The *KBAgent* uses the same reasoning mechanism as suggested by Lin et al. [12] for identifying an opponent's type. This mechanism is based on the Bayesian updating rule that updates or revises beliefs in light of new evidence a posteriori ([9], Chap. 2). As incomplete information in the environment is modeled by the existence of several agent types in each time period, the agent consults the component in order to update its belief regarding the other party's type.

Recall that there are k possible types of agents. At time $t = 0$ the prior probability of each type is equal, that is, $P(\text{type}^i_{t=0}) = \frac{1}{k}, \forall\, i \in$ **Types**. For each time

[1] In order to perform this derivation we replace α_t with $u_{KB}(\beta_t, t)$ where β_t is the threshold agreement, that is all offers with utility values higher than the utility if it were accepted. We then take derivative by β_t.

period t we calculate the *a* posteriori probability for each of the possible types, taking into account the history of the current negotiation. This is done *incrementally* after each offer is received or accepted. That is, the believed type is updated every time an offer is received or accepted, thus eventually it is based on the entire history thus far. Thereafter, this value is assigned to P(type$_t$). Using the calculated probabilities, the agent selects the type whose probability is the highest and proposes an offer as if it were the opponent's type. Formally, at each time period $t \in$ **Time** and for each type $\in$ **Types** and $\vec{o}_t \in O$ (the offer at time period t) we compute:

$$P(\text{type}^i|\vec{o}_t) = \frac{\mathrm{P}(\vec{o}_t|\text{type}^i)\mathrm{P}(\text{type}^i_t)}{\mathrm{P}(\vec{o}_t)} \tag{8}$$

where $\mathrm{P}(\vec{o}_t) = \sum_{i=1}^{k} \mathrm{P}(\vec{o}_t|\text{type}^i) \cdot \mathrm{P}(\text{type}^i_t)$.

Now we can deduce the believed type of the other party for each time period t, $BT(t)$, using the following equation:

$$BT(t) = \arg\max_{i \in \mathbf{Types}} \mathrm{P}(\text{type}^i|\vec{o}_t), \quad \forall t \in \mathbf{Time} \tag{9}$$

3 The Colored-Trail Environment

The Colored Trails (CT) game [7] is a general negotiation test bed that provides an analogy to task settings in the real world.[2] CT is played on a board of colored squares. Players are issued colored chips and are required to move from their initial square to a designated goal square. To move to an adjacent square, a player must turn in a chip of the same color as the square. Players must negotiate with each other to obtain chips needed to reach the goal square (see Fig. 2). Each participant obtains a score at the end of the game that is computed as follows: X points are given for reaching the goal square and Y bonus points are given for each chip that each agent retains at the end of the game. If the player did not reach the goal, Z penalty points are given for each square between the participant's final position and the goal square. Note that in this game, the performance of the agent does not depend on the outcome of the other player. Agreements are not enforceable, allowing players to promise chips but not transfer them. In addition, each player can see the entire game board. This environment is interesting for people to play as it is analogous to task settings and has a vivid representation of strategy space (as opposed to just a list of outcomes in the other environment).

The simulation environment is adaptable such that different variations of the game can be set. The size of the board, the number and color of total chips, and chips given to each player can be changed. In addition, the information accessible to each player can

[2] Colored Trails is Free Software and can be downloaded at http://www.eecs.harvard.edu/ai/ct

Fig. 2 Snapshots of Colored Trails GUI

also be controlled (e.g., whether the player has full information of the others player's chips). The automated agents can play both sides in the game, while the human counterpart accesses the game via a web address. The game itself is split into turns, where each turn is divided into a negotiation phase, a transfer phase and a movement phase. In the negotiation phase, the players can request or promise to send chips. Then, in the transfer phase, the players choose the extent to which they meet their commitments by deciding the amount of chips to send. This might result in one agent sending the promised chips in return for the other agent's chips, while the other agent fails to deliver. In the movement phase, the players can choose to move to adjacent squares, given that they have the required colored chips. The game terminates when either side reaches the goal square or if any of the players has not moved in three consecutive turns. We continue to describe the *PURB* agent, which was designed and evaluated in this environment using different variations of the game (board size, chips, dependency on the other player for reaching the goal square) [5, 6].

3.1 The PURB *Agent*

The *PURB* agent stands for Personality and Utility Rule Based agent. It combines a social utility function representing the behavioral traits of other participants with a rule-based mechanism using the utility function to make decisions in the negotiation process. This agent was designed to model and adapt its behavior to the individual traits exhibited by its negotiation partner by making trade offs between acting selfishly to achieve short-term benefits and exhibiting helpful behavior towards maintaining long-term relationships. This is achieved by the agent dynamically changing its proclivity towards cooperation and reliability and also modeling the same traits in its opponent. A utility function is used to evaluate each possible action and proposal, and randomization is used to choose between different options [6]. As opposed to the *KBagent* described in Sect. 2.1, the *PURB* agent did not have access to previous negotiation data. It was therefore implemented with a set of rules which allow it to adapt its behavior based on current game play.

The agent negotiated with human subjects in the United States and Lebanon. Results showed that the agent was able to adopt a different negotiation strategy for each country. Its average performance across both countries was equal to that of people. However, the agent outperformed people in the United States because it learned to make offers that were likely to be accepted, and that were more beneficial to the agent than to people; the agent was outperformed by people in Lebanon, because it adopted a high reliability measure which allowed people to take advantage of it.

Utility based strategy: The *PURB* agent modeled its negotiation partner in terms of two behavioral traits: The *helpfulness* trait of an agent described the willingness of negotiators to share resources with others. This was measured as the percentage of proposals in a game in which more chips were offered to the human than to the agent. The *reliability* trait of an agent described the degree to which negotiators kept their commitments. This was measured as the ratio between the number of chips transferred by a player after an agreement and the number of chips that were promised by the player. Together, these behavioral traits represented the *cooperativeness* measure of a negotiator.

The social utility functions that the *PURB* agent employed are a weighted aggregate of features relating to its potential benefit from potential actions, and how these actions affected the way the human and the agent reasoned about each other's reliability and helpfulness over time.

For the remainder of this section we will use index i to denote the *PURB* agent and index j to denote the person.

For any state s that encompasses a CT board, players' locations and chip settings, the social utility of the *PURB* agent, denoted u_i, was a function that was a weighted combination of the following features:

1. The expected future score for agent i. This score was computed using a heuristic that underestimated the benefit to the agent from a potential exchange in future rounds, given that the agent does not give any chips away. It was estimated as follows: Let $P_i(G / s, p)$ denote the probability that i will get to the goal given that proposal p is fulfilled. The expected future score was defined as

$$(P_i(G|s,p) \cdot 100) + ((1 - P_i(G|s,p) \cdot 100) \cdot 10 \cdot d) + s \cdot 5$$

where 100 is the number of bonus points to get to the goal according to the CT scoring function; d is the Manhattan distance of i from the position on the board given that the exchange was fulfilled and the goal square; 10 is the number of penalty points for each square in the distance from the final position of *PURB* and the goal square; s is the number of chips left in the player's possession after it advances to the goal using the shortest possible path, and 5 is the number of points awarded to the player for each chip left in its possession at the end of the game.

2. The expected future score for agent j (computed in the same way as for i).
3. The cooperativeness measure of agent j (in terms of helpfulness and reliability). This feature represented i's beliefs of j's model of the reliability and helpfulness of j.
4. The cooperativeness measure of agent i (in terms of helpfulness and reliability). This feature represented i's beliefs about how j perceives its own measure of cooperativeness.

The weight of each aforementioned feature was established by a set of predetermined rules, and relied on the dependency relationships between participants as well as their cooperative measures.

Each time an agreement was reached and transfers were made in the game, i updated the helpfulness and reliability measures of both agents (these values were aggregated over time using a discounting rate of 0.1).

Using this social utility allows the *PURB* agent to vary its strategy based on its estimate of the other participant's cooperativeness measure. For example, if the reliability of the other participant was high, this would increase the social utility of actions that favor the other participant.

The *PURB* agent also applied a set of rules that narrowed the search space of possible actions to be considered by the agent's utility function. These rules depended on aspects relating to the state of the game (e.g., the number of chips each agent had, whether a participant can independently reach the goal). At each step of the game, the agent used its social utility function to choose the best action out of the set of possible actions that were constrained by the rules. The rules were designed such that the *PURB* agent begins by acting reliably, and adapts over time to the individual measure of cooperativeness that is exhibited by its negotiation partner.

Generating proposals: The *PURB* agent generated a subset of possible offers based on its utility function. It non-deterministically chose any proposal out of the subset that provided a maximal benefit (within an epsilon interval) according to the agent's social utility function. A set of rules were applied to facilitate the agent's decision-making by reducing the set of possible proposals to consider when using its utility function.

Accepting proposals: The *PURB* agent accepted an offer if it was more advantageous to it than the offer it would make as a proposer in the same game state, or if accepting the offer was necessary to prevent the game from terminating. Formally, let $u_i(p_j, \text{accept} \mid s)$ denote the utility for i from an offer p_j made by j at state s. Let p_i denote the offer that agent i would make at state s. Generally, agent i would accept an offer p_j if $u_i(p_j, \text{accept} \mid s) \geq u_i(p_i, \text{accept} \mid s)$. In addition, i would accept any proposal that prevented the game from ending, which occurs when the following conditions hold: (1) the chips in the possession of agent i do not allow it to move on the board at state s; (2) the offer p_j allows agent i to move; and (3) if i rejects the offer, the limit for dormant turns will be reached and the game would end.

Lastly, the degree to which *i* accepted proposals from *j* that made *j* task-independent (i.e., they allowed *j* to get to the goal) but where *i* would remain task-dependent occurred only when *i* was very certain that *j*'s reliability was high. The degree of certainty of *i* regarding the reliability of *j* grew in proportion to the number of observations of *j*'s behavior.

Transferring chips: These rules specify the extent to which the *PURB* agent fulfilled its agreements in the game. This behavior directly depended on its model of the other's cooperativeness. If the reliability of *j* was *low*, it was likely that the other participants would not fulfill their agreement. Therefore *i* did not send any of its promised chips. If the reliability of *j* was *high*, then *i* sent all of the promised chips. If the reliability of *j* was *medium*, then the extent to which *i* was reliable was directly affected by the dependency relationships in the game:

1. If *j* was *task dependent*, and the agreement will result in *j* becoming task *independent*, then *i* sent the largest set of chips such that *j* remained task dependent.
2. If the exchange resulted in *both* agents becoming task independent, there was a high likelihood that *j* will fulfil the agreement. In this case, *i* sent all of the promised chips. Otherwise, *i* sent half of the promised chips with a given probability, because it was not certain that *j* would fulfil its agreements.
3. Lastly, if the exchange resulted in both agents being task dependent, then *i* randomized between fulfilling its agreement and sending all of the promised chips, or sending a subset of the chips. Again, this is because *i* was not certain that a participant with a medium measure of cooperativeness would meet its commitment.

After the general description of the *KBAgent* (Sect. 2.1) and the *PURB* agent (Sect. 3.1), we describe in the next section how we employed a novel approach to evaluate the proficiency of these agents and assist in the process of their evaluation with people.

4 Facilitating Agents' Evaluation

Using people for the evaluation purposes is costly in terms of time, effort and money, making it a difficult task for researchers. The traditional view was that people cannot be replaced in the evaluation process because of their diverse behavior, which is affected by cognitive, social and cultural factors [2, 8].

In Ref. [10], we investigated the use of Peer Designed Agents (PDA) – computer agents developed by human subjects – as a tool for facilitating the evaluation process of automatic negotiators that were developed by researchers. We suggested a novel approach for evaluating the efficacy of automated agents by evaluating them with PDAs and not people. Thus, automated agents can be first evaluated and improved before moving on to experimentations with people.

This paradigm was inspired by the "strategy method" commonly used in behavioral economics, which is an experimental methodology that elicits from people a complete strategy specification. People state their actions for every possible situation that may arise in their interaction [15, 18, 19]. The assumption behind this method is that people are able to effectively encapsulate their own strategies if they are properly motivated, monetarily or otherwise.

In several different experiments we compared the performance of the *KBAgent* against people and against PDAs in the GENIUS environment, and the performance of the *PURB* agent with people and PDAs in the *Colored-Trail* game. The experiments included more than 300 human subjects and 50 PDAs developed by students.

Results show that PDAs can be used as a mechanism for evaluating automated negotiators that reflects on the behavior of people, as well as to fine-tune and improve the strategy of the automated negotiators. In particular, the *KBAgent* and the *PURB* agent were able to outperform people in our experiments in the same situations in which they were able to outperform PDAs. In addition, the results showed that the automated agents exhibited the same measure of generosity towards their negotiation partners. These patterns occurred for all types of domains, and for all types of automated negotiators, despite the fact that there were individual differences between the behavior of PDAs and people. These results provide an empirical proof that PDAs can be used to better determine the proficiency of an automated negotiator when matched with people, as well as to compare between the performance of different automated negotiators and their effects on human behavior. Thus, they can alleviate the evaluation process of automatic negotiators, and facilitate their design.

There are fundamental benefits to using PDAs instead of people. First, PDAs are accessible 24/7 and can be used whenever needed. In addition, PDAs are not biased and thus can be used several times to asses the automated agent's behavior. Thus they allow the agent designer to revise and change her agent with the ability to evaluate each design and compare it to previous designs. Lastly, it allows different automated agents to be matched on the same set of PDAs and obtain an objective evaluation of the results.

5 Conclusions

The importance of designing proficient automated negotiators to negotiate with people and evaluating them cannot be overstated. Yet, evaluating agents against people is a tiresome task, due to the cost and time required. In this paper we presented two infrastructures for designing agents and an overview of results of an extensive systematic experimentation. Our results revealed that playing well against peer designed agents can reflect on the proficiency of the automated negotiator when matched with people. Moreover, we showed that while PDAs result with different negotiation outcomes than people, there is a common behavioral pattern when they are both matched with automated agents.

The two infrastructures we presented – the GENIUS environment [11] and the Colored Trails game environment [7] – can be used for general purpose research on bargaining with people and automated agents, each with its distinct features.

The GENIUS environment allows finite horizon bilateral negotiations between two people or agents. It allows the incorporation of any domain, with or without utility attached to it, and with or without a time discount factor. It also allows incorporation of mediator and statistical tools to assist during the negotiation. It basically follows an alternating offer protocol and can allow for unenforceable agreements. In addition it allows for complete or incomplete information, when incomplete information is expressed as uncertainty regarding the utility preferences of the opponent, and it is assumed that there is a finite set of different negotiator types. The second environment of the CT game is a general negotiation test-bed that provides an analogy to task-settings in the real world. Its vivid representation of strategy space (as compared to just a list of outcomes in the GENIUS environment) makes it more interesting for people to play. Like the GENIUS environment, automated agents can also play the game, and researchers can vary the different settings and complexity of the game (e.g., repeated vs. one-shot, availability of information, degree of dependency and communication protocol).

The two automated agents we described are also distinct from one another, not only by the environment in which each was used. The *KBAgent* is a domain independent agent, which implements several tactics and heuristics that are qualitative in nature. The main difference between it and the *PURB* agent is that it had access to data from previous interactions and used it in its decision-making component to increase its proficiency. It also used non-deterministic behavior via means of randomization. The *PURB* agent, on the other hand, is suited to be used in the CT game environment alone. Its non-deterministic behavior is based on rules which allow it to be adapted to emergent situations of the game, as it does not have access to past game data.

We have shown that it is indeed possible to develop automated agents that interact proficiently with people. The availability of generic infrastructures and the possibility of evaluating automated agents with peer designed agents to evaluate their performance with people should motivate more and more researchers to pursue the design of automated agents. Some of the drawbacks of designing automated negotiators that are capable of negotiating with people stems from the fact that their evaluation involves people and the cumbersome task of orchestrating the experiments with people. Perhaps simplifying this evaluation process could make the design of these agents more accessible and motivate researchers to design more automated agents that are specifically designed to negotiate proficiently with people.

References

1. Bazerman, M.H., Neale, M.A.: Negotiator rationality and negotiator cognition: The interactive roles of prescriptive and descriptive research. In: Young, H.P. (ed.) Negotiation Analysis, pp. 109–130. The University of Michigan Press (1992)
2. Camerer, C.F.: In: Behavioral Game Theory. Experiments in Strategic Interaction, chap. 2. Princeton University Press (2003)
3. Davis, R., Smith, R.G.: Negotiation as a metaphor for distributed problem solving. Artificial Intelligence 20, 63–109 (1983)

4. Erev, I., Roth, A.: Predicting how people play games: Reinforcement learning in experimental games with unique, mixed strategy equilibrium. American Economic Review 88(4), 848–881 (1998)
5. Gal, Y., Grosz, B., Kraus, S., Pfefer, A., Shieber, S.: Agent decision-making in open mixed networks. AIJ (accepted, 2010)
6. Gal, Y., Kraus, S., Gelfand, M., Blumberg, Y., Salmon, E.: An adaptive agent for negotiating with people in different cultures. ACM Transactions on Intelligent Systems and Technology (ACM TIST) (accepted, 2010)
7. Grosz, B., Kraus, S., Talman, S., Stossel, B.: The influence of social dependencies on decision-making: Initial investigations with a new game. In: AAMAS. pp. 782–789 (2004)
8. Lax, D.A., Sebenius, J.K.: Thinking coalitionally: party arithmetic, process opportunism, and strategic sequencing. In: Young, H.P. (ed.) Negotiation Analysis, pp. 153–193. The University of Michigan Press (1992)
9. Leonard, T., Hsu, J.S.J.: Bayesian Methods – An Analysis for Statisticians and interdisciplinary Researchers. Cambridge University Press, Cambridge, UK (1999)
10. Lin, R., Kraus, S., Oshrat, Y., Gal, Y.K.: Facilitating the evaluation of automated negotiators using peer designed agents. In: AAAI. pp. 817–822 (2010)
11. Lin, R., Kraus, S., Tykhonov, D., Hindriks, K., Jonker, C.M.: Supporting the design of general automated negotiators. In: ACAN (2009)
12. Lin, R., Kraus, S., Wilkenfeld, J., Barry, J.: Negotiating with bounded rational agents in environments with incomplete information using an automated agent. AIJ 172(6–7), 823–851 (2008)
13. Luce, R.D.: Individual Choice Behavior: A Theoretical Analysis. John Wiley & Sons, NY (1959)
14. McKelvey, R.D., Palfrey, T.R.: An experimental study of the centipede game. Econometrica 60(4), 803–836 (1992)
15. Offerman, T., Potters, J., Verbon, H.A.A.: Cooperation in an overlapping generations experiment. Games and Economic Behavior 36(2), 264–275 (2001)
16. Osborne, M.J., Rubinstein, A.: A Course In Game Theory. MIT Press, Cambridge MA (1994)
17. Oshrat, Y., Lin, R., Kraus, S.: Facing the challenge of human-agent negotiations via effective general opponent modeling. In: AAMAS. pp. 377–384 (2009)
18. Selten, R., Abbink, K., Buchta, J., Sadrieh, A.: How to play (3×3)-games: A strategy method experiment. Games and Economic Behavior 45(1), 19–37 (2003)
19. Selten, R., Mitzkewitz, M., Uhlich, G.R.: Duopoly strategies programmed by experienced players. Econometrica 65(3), 517–556 (1997)
20. Tversky, A., Kahneman, D.: The framing of decisions and the psychology of choice. Science 211, 453–458 (1981)
21. Wand, M., Jones., M.: Kernel Smoothing. Chapman & Hall, London (1995)

Virtual Technologies and Empowerment of Users of Rehabilitation

Naomi Schreuer and Patrice L. (Tamar) Weiss

Abstract Empowerment is a multi-dimensional social process that helps people gain control over their own lives. Empowerment of people with disabilities entails the motivation to control as well as the knowledge and skills to adapt to and influence one's own rehabilitation process. The goal of empowered rehabilitation is to guide users to achieve their maximal independence by developing the skills necessary to overcome the physical, social and emotional barriers in their lives. We are currently witnessing a profound change in social attitudes where, after years of repression, people with disabilities struggle for their human rights and for control over resources. Information and communication technologies (ICT) have been used to empower populations who are weak due to gender, poverty, rural residence, age and disability. In this chapter, we have applied a model of patterns of social change to show how the use of novel ICT technologies may contribute to social change and empowerment of people with disabilities. Each of the model's four patterns – social planning, social reform, social development and social action – is illustrated with an example of the role of technology during this process. Lessons learned during the empowerment process for people with disabilities may also be incorporated into general ICT programs to help enhance usage in additional segments of the population including minority groups as well as those who have traditionally been challenged in this area (e.g. the elderly, people who live in remote locations).

1 Introduction

Despite a widespread shift from the traditional medical model to a social model of disability, the persistence of medical and welfare approaches to disability still prevail [1]. In most countries around the world, persons with disabilities are still

N. Schreuer (✉) • P.L. (Tamar) Weiss
Department of Occupational Therapy, University of Haifa, Mount Carmel, Haifa, Israel
e-mail: schreuer@research.haifa.ac.il; tamar@research.haifa.ac.il

A. Krüger and T. Kuflik (eds.), *Ubiquitous Display Environments*,
Cognitive Technologies, DOI 10.1007/978-3-642-27663-7_13,

© Springer-Verlag Berlin Heidelberg 2012

regarded as 'sick', and therefore unable to cope as others do. One of the major obstacles that they encounter is social inaccessibility, i.e., difficulty in integrating into the general community [2]. According to the social model of disability and human rights movement, the participation of persons with disabilities in political and legislative decisions concerning their lives is essential [1]. Inclusion should be regarded as a mutual process wherein persons with disabilities are empowered to take an active part in changing their social status and, ultimately, to lead the process [3, 4].

In recent years, a process of growing awareness, technological developments and increased activity among persons with disabilities, have gained momentum opening new opportunities for their full inclusion into society. Nevertheless, they continue to experience numerous difficulties in fulfilling their potential in education, employment and social interactions. Recent legislation has endeavored to meet their needs via, for example, laws to make public buildings and ICT more accessible [5], as well as mainstream children with special needs into regular school programs. The social change described may occur merely with the presence of empowerment, which viewed hereby as means to develop the skills necessary to overcome the physical, social and emotional experienced by people with disabilities.

2 Empowerment

The verb 'to empower' was first used in the seventeenth century to denote 'authorize', 'delegate', or 'enable' within the context of legislative control. The twentieth century embraced the notion of empowerment in discussions related to political, psychological and business applications [6]. Here is a widely accepted definition of empowerment is:

> a multi-dimensional social process that helps people gain control over their own lives. It is a process that fosters power in people for use in their own lives, their communities and in their society, by acting on issues they define as important [7, p. 1].

This definition assumes that individuals have a basic need for some level of control during interactions with the environment. An individual's power during these interactions is defined as "the capacity to implement" [8]. The broad meaning of power encompasses domination, authority, influence, collaboration, sharing and mutuality [8]. Empowerment is possible only if power can create meaningful change; it depends upon the idea that power can expand, and occurs at various levels, individual, group, and community [9].

The importance of empowerment was emphasized by studies that showed a positive correlation between it and psycho-social factors such as quality of life and self-esteem [10, 11]. Studies also validated empowerment as a universal phenomenon which does not depend on specific characteristics of people, such as age, gender, race and marital status [10, 12].

In the political context, theorists and policy makers have recognized that change at the level of the individual is a prerequisite for collective change to occur at both community and societal levels. The synthesis between individual and collective change is that empowerment enables individuals to become partners in solving complex issues in a creative and realistic manner via collaboration based on mutual respect and vision that includes diverse perspectives [13]. In the socio-political arena, empowered individuals are critically involved and able to identify what must change, as well as participate in the decision-making that leads to behavioral change [14]. In the business management sphere, a large body of knowledge evolved from organizations that adopted empowerment initiatives, focusing on the relationship between empowerment and organizational performance [6]. Spreitzer [16] found six significant social elements that explain 70% of the variance of an individual's empowerment at work: role ambiguity; working for a boss who has a wide range of control; sociopolitical support; access to information; access to resources; and participative climate. These elements also reflect major components of empowerment in arenas other than business.

Rappoport [9] noted that it is easier to define empowerment by its absence than in action since it is a concept that takes on various forms for different people and in diverse contexts. Moreover, the appraisal of empowerment and the factors that enhance it often depend on the investigator's discipline as well as personal biases since it challenges one's basic assumptions about power, helping, achieving, and succeeding [7].

3 Empowerment, Technology and Rehabilitation

The business arena has identified the potential of technology as a tool to increase employees' empowerment. The increased use of technology, especially in the work force, has led to the term "user-empowerment", relating to usage of information and communication technology (ICT) [6]. User-empowerment in computer mediated environments is a distinct term within the self-empowerment notion, relating to user autonomy, computer self-efficacy, intrinsic motivation, and perceived usefulness [16]. ICT enables user-empowerment by enhancing access to information, augmenting operational and decisional activities, improving individual performance, and contributing to overall organizational performance [6].

The Internet is one of the most powerful ICT tools that is used to empower individuals [17]. Termed online or e-empowerment, it serves to break barriers by reframing identities, increasing self efficacy and skills, and compensating for social inadequacies via group reinforcement [18]. For people with disabilities, ICT and other technologies have even greater potential to increase their opportunities by obviating the need for manual and mobility skills. For example, computerized accessibility adaptations enable access to all forms of communication that would otherwise be impossible for a person who is visually or movement impaired [19].

On the other hand, the operation of ICT entails a certain level of cognitive and perceptual ability, training, equipment and access to other resources [19].

The implications of the digital gap for people with disabilities have been demonstrated via analysis of Internet diffusion in ten European countries [20]. This study revealed that disability status has a significant negative effect on the probability of being an Internet user, even for those in higher income brackets and with greater educational levels (individuals without disabilities are 1.2 times more likely to use the Internet than those with disabilities). Interestingly, they found that the current gender related digital gap (i.e., men are 1.5 times more likely to use the Internet than women) is larger than the disability related gap. However, among those who do use the Internet, no significant relationship was found between disability and frequency of use. These findings are similar to American findings [21].

Vicente and López [20] describe a vicious cycle between ICT and disability. Individuals with disabilities are much less likely to be included in fully paid employment, which limits their access to and ability to fund ICT. They lack skills and training, especially due to primary and secondary learning disabilities [22], and also face many accessibility problems. For example, only 5% of web sites meet the accessibility standards [23]. All these factors detract from ICT usage which, in turn, leads difficulty in gaining the necessary digital skills in order to obtain skilled employment and fuller inclusion in society.

Empowerment of people with disabilities entails the motivation to control as well as the knowledge and skills to adapt to and influence one's own rehabilitation process [24]. This approach underlines the importance of participation and control, and the view of wellness versus illness and competence versus deficiency. The goal of empowered rehabilitation is to guide users to achieve their maximal independence, by developing the skills necessary to overcome the physical, social and emotional barriers in their lives [25].

We are currently witnessing a profound change in social attitudes where, after years of repression, people with disabilities struggle for their human rights and for control over resources [26]. They are still widely assumed to be unable to know their own needs or to act on them. They are viewed as a stigma, emanating from a paternalistic stance that limits the number and type of decisions they are "permitted" to make. These attitudes together with a pervasive lack of practice, preserves their dependence in the long run, unless they are given opportunities to make important decisions [27]. People with disabilities tend to lack access to information, which supports professionals' claim of their impulsivity, irresponsibility or manipulative behavior during decision making. A vicious circle with results reinforcing prejudices ensues. Empowerment is a means to end up this circle by giving people with disabilities opportunities to adopt the notion that they have rights and that they can create change [27]. It will be achieved via access to information and resources, making decisions and choices out of options, critical thinking, assertiveness in using one's rights to create change, feeling part of a powerful group, and self-esteem in overcoming stigma [28].

ICT has been used to empower populations who are weak due to gender, poverty, rural residence, age and disability [29, 30]. In recent years, virtual reality (VR)

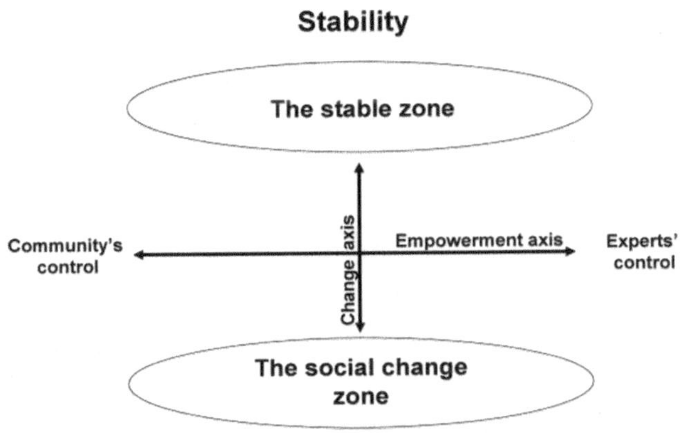

Fig. 1 Social change and empowerment as presented by Jeffries [37]

technologies have begun to be used as an assessment and an intervention tool in rehabilitation [31–34]. Virtual environments provide people with disabilities with safe access to interactive and realistic situations that would otherwise be inaccessible to them due to their motor, cognitive or psychological limitations [35]. The ability to change the virtual environment relatively easily, to grade task difficulty, and to adapt it according to the user's capabilities are important assets of VR [32]. Moreover, people who have experienced the use of virtual environments have reported the activity to be enjoyable, motivating, and empowering [26, 32–36].

4 Model for Social Change Based on Empowerment

In an effort to better understand technology's potential to enhance the empowerment of people with disabilities, we apply a theoretical model of social change in this section. We then present examples to illustrate the relevance of the model's four patterns of change.

Jeffries [38] developed a model to create a social change for individuals and groups with disabilities. The model aims to change attitudes, structures, policy, support and/or redistribution of resources. In order to create change, every organization or group should examine the orientation of its services along the two axes shown in Fig. 1; the vertical (change) axis depicts a continuum from stability to change, and the horizontal (empowerment) axis depicts a continuum from professional (expert) dominance to control by individuals or community groups.

The empowerment axis describes the daily practice and tension between control and balance of power that professionals and clients with disabilities experience during the decision making process involving both services and resources [38]. Churchman and Barak [39] used Jeffries' [38] model to analyze the forces that drive social change for people with disabilities, with a particular focus on the situation in Israel. They divided the graph into four major patterns of change, one per quadrant,

Fig. 2 Model of social change patterns for people with disabilities (Adapted from Churchman and Barak [39])

as shown in Fig. 2. We have applied their adapted model to illustrate how ICT, virtual reality and other novel technologies may contribute to social change and empowerment of people with disabilities. All four patterns are required to achieve a real change from the stable, passive status of people with disabilities to one of empowerment as typified by their involvement and participation in individual activities and societal change [39].

5 Patterns of Change

5.1 Social Planning

Plans and activities in this quadrant (upper, right section of graph in Fig. 2), focus on planning carried out primarily by experts with input from grassroots community groups. The role of this input is to assess community needs and to seek the appropriate program to respond to those needs [39]. In most cases, organizations prefer to stay in this safe, stable zone where change occurs only in improving the coordination of existing services in favor of people with disabilities [40]. Virtual reality technologies may contribute to social planning by improving communication and sharing information. Long distance communication, discussion forums, and online surveys are good examples for this contribution. Virtual technology may make a real difference where in most cases experts intend to stay in the safe, stable zone, where the community members gain knowledge, skills, and a small amount of empowerment towards the achievement of change.

We illustrate the social planning quadrant with an example of how technology has been used to support people with intellectual and developmental disabilities (IDD). These individuals are known to be particularly dependent in terms of both physical and social activities [40, 41]. Indeed, the low level of fitness found in persons with IDD is attributed to their passive lifestyle [42, 43], low motivation [44] as well as other psychological and physiological factors [45]. There is an ever-growing awareness of the importance of participation in leisure activities as a crucial component of empowerment and quality of life among people with IDD

[46–48]. However, the degree of participation in active leisure activities by people with severe cognitive limitations is often extremely limited [49–52]. These findings are often attributed to the limited range of adapted activities [50], and the intensive support required by this population in order for them to participate in leisure activities [51]. Other barriers to participation in leisure for individuals with disabilities include inaccessibility to facilities, transportation limitations, and apprehension on the part of the caregivers [53]. The majority of pastimes available to this population still tend to involve sedentary activities such as watching television and modified arts and crafts activities [53].

Advanced technologies which are adaptable address their need for increased exposure to augmented leisure opportunities and documents the positive correlation between life satisfaction, self-esteem, companionship, enjoyment and relaxation and leisure satisfaction [54]. The pervasive lack of opportunity for a wide choice of independent leisure time activities may contribute to the development of dependent behavioral patterns, learned helplessness and depression [54–57]. Provision of ICT that enhances leisure opportunities as well as participation in leisure activities contributes to an improvement in self-confidence and self-concept amongst individuals with cognitive limitations, an improvement in social and motor skills, increased communication abilities, greater physical fitness and weight reductions [58–60].

For the past decade, VR has been used to help people with IDD to achieve daily living skills among higher functioning participants (cf. a review [61]). In a series of three companion studies, we have shown that independent leisure experiences such as those provided within video capture virtual environments [62–64] expanded the repertoire of avocational opportunities available to adults with severe intellectual and physical disabilities; participants with mild to severe levels of IDD and physical limitations were tested. They experienced a VR intervention program using different types of video capture technology operated by an occupational therapist with expertise in technology; in each study, a control group of residents at a similar level performed traditional activities.

A limitation of the first two studies [62, 63] was that VR service delivery was made available only by research funded personnel. The crucial elements to participation in exercise programs are: environmental availability, involvement of on-site support personnel and a highly motivating yet feasible activity [65–67]. It was therefore necessary to identify a way to imbed the VR activity within the daily routine of a residential setting with involvement of in-house staff. In the third study [64], we demonstrated that similar effects were repeated by training the in-house caregiving staff to independently implement a VR game-based exercise program.

The use of VR for IDD shifted systematically from a laboratory driven research setting to a community residential setting by training the in-house caregivers to engage the end users in games and activities to enhance their participation and to enrich their experiences. The caregivers functioned as mediators to use the new technologies to improve the rehabilitation outcomes and empower the target population. Nevertheless, both caregivers and people with IDD lacked knowledge and control over the technology and activities and were dependent on VR experts.

5.2 Social Development

Plans and activities in this quadrant (upper, left section of graph in Fig. 2), relate to classical community development, which focuses on developing the potential of and opportunities for its members. However, social development may involve members in new projects that are within the consensus, without including them in large scale decision-making [38, 39]. Virtual reality technologies may contribute to social development by enhancing the six key elements of empowerment identified by Chamberlin [68]: access to information and resources, making decisions and choices out of options, thinking critically, assertiveness in using rights and creating change, feeling part of a powerful group, and self-esteem overcoming stigma.

The rapidly increasing access to remote medical and rehabilitation expertise via tele-health is an illustrative example of the social development pattern. The primary impetus behind the development and provision of tele-health services is the desire to provide equitable access to rehabilitation services for individuals who are remote from specialists, or those who have limited access as a result of a physical disability which prevents them from attending a local service [69]. This is often the case for people who have suffered a stroke, traumatic brain injury, or pediatric developmental disorders such as Autism Spectrum Disorders (ASD). Other driving factors for tele-health services include the potential transportation cost and time savings that may be enjoyed by both the health care system and the client, the continuity of patient care that can be achieved through the remote provision of services, the ability to better control the timing, intensity and sequencing of interventions, and other intangible benefits such as the positive effects of providing care in a client's own social and vocational environment [70].

Employing tele-health technologies for home use enables fundamental advances in the way health care is provided due to the provision of continued care and education after a client has been discharged from hospital, advancement of the quality and efficiency of health-care delivery, and access to outcome data of procedures and evaluation results [71]. Client self-care and self-management, activities that individuals perform to maintain health and manage illness, had, traditionally, been under the control of the medical caregiving staff. Transfer of much of this responsibility to the individual is thought to empower them to become more proactive in their own care and to assume a greater share in decision-making [72]. These activities appear to be particularly suited to home-based tele-health programs [73].

For example, the SMART rehabilitation project studied the use of a movement tracking system (two small inertial sensor attached to the client's arm), a personal computer and a webserver to facilitate the self-management of functional tasks to motivate stroke patients to exercise their upper limb while at home [74]. The client attaches the sensors, selects a specific functional goal such as drinking or reaching, and repeats the task a number of times. The sensors record changes in the client's arm position resulting from movement of the elbow and shoulder joints. The information collected by the sensors is analyzed and provided as feedback to the client (and caregiver) in the form of on-screen 3D avatars showing the client's movement compared to the targeted "normal" movement. Summative kinematic data on frequency of use, angle ranges, and cycle time can also be viewed.

These sensor-based technologies monitor performance and provide feedback on progress, display educational guidelines, and alert both client and professionals to potential risks [71]. The results demonstrated the feasibility of using ICT and sensor technology to enable the key elements of the self management process to take place.

The social development quadrant focuses on fostering the potential of and opportunities for its members. The process is novel; for example, tele-health uses new technologies to enable the provision of conventional rehabilitation in a person's home. However, the content is quite traditional (a therapist designed exercise program) that is formulated and implemented with minimal inclusion of the end user in decision-making.

5.3 Social Reform

Plans and activities in this quadrant (lower, right section of graph in Fig. 2), focus on making structural and political changes for people with disabilities, but led by the experts and their agenda, rather then by the community and its leadership [38]. Experts advocate people with disabilities, but leave them behind, expecting them to follow and comply with the reform. Virtual technologies may contribute to psychological empowerment in the workplace by facilitating an active, rather than a passive orientation toward a work role defined by the employer [15].

We illustrate the social reform quadrant with e-mentoring technologies. E-mentoring entails the use of electronic media (e.g., email, chat) by a mentor to provide support, information and social interaction to an individual in need (referred to as a protégé). There are a number of e-mentoring programs available for people with disabilities [75] however these typically have a person without a disability in the role of the mentor. Social reform occurs when protégés are mentored by people who also have a disability. For example, DO-IT (http://www.washington.edu/doit) is an e-mentoring program that matches mentors and protégés with similar disabilities in order to encourage the latter to participate in postsecondary academic programs. The objective is to help them identify appropriate careers via disability awareness training and information dissemination. The protégés are high school students with disabilities (hearing, mobility, vision, and health impairments) who are interested in science, engineering, mathematics, and technology. They use computers, assistive technology, and the Internet to communicate with each other and with the mentors who also have disabilities and are currently enrolled in postsecondary institutions or are employed [75]. MENTOR, the American national mentoring partnership, established an e-mentoring program designed to promote successful transition of youth with disabilities to adult life (http://www.mentoring.org; http://ici.umn.edu/ementoring). This program facilitates interactions between a student with a disability and an employer focuses on building a career path. Another e-mentoring program which matches youth and adults with disabilities is conducted under the auspices of the Partners for Youth Disabilities [76]. Shpigelman et al. [77] demonstrated that mentor-protégé pairings where both partners have disabilities could be highly effective in providing informational support, tangible assistance, social support, or emotional support.

Studies of the effectiveness of e-mentoring programs suggest that individuals who are isolated, stigmatized, or lack real-world social support may be especially motivated to participate in and gain from these online support interventions [78–80]. They interviewed women with somatic diseases (breast cancer, fibromyalgia, or arthritis) who participated in online support groups, in order to explore if, and in which ways, they felt empowered by their participation. Online support groups appeared to make a valuable contribution to the emergence of empowered persons with disabilities. The participants were being better informed, felt confident in the relationship with their physicians, their treatment, and their social environment. They also achieved greater acceptance of the disease, and increased their optimism, control, self-esteem and social well-being. Bowker and Tuffin [81] found that online experiences afford people with disabilities an opportunity to enjoy greater social acceptability and self-esteem, independent of their identity as a person with disability.

5.4 Social Action

Plans and activities in this quadrant (lower, left section of graph in Fig. 2), focus on making changes so that a new relationship occurs between participants and social institutions [38]. Minorities such as people with disabilities struggle in social actions for real participation in political decisions, to enhance services and redistribution of resources: "nothing about us, without us" [27]. As a community's members become more and more involved in the process of change, the experts withdraw gradually, until they relinquish control [38]. Virtual technologies which empowers workers in enterprise systems, may serve social action of minorities by enhancing their operational and decisional activities, improve their individual performance metrics, and contribute to the overall organizational performance (i.e., by gathering information, coordinating and distributing their message).

Second Life is a 3D virtual world that enables participants to create identities in the form of avatars that interact with others in an immersive environment (http://lindenlab.com). Users of Second Life engage in social interactions, shop, run businesses, and access a variety of information and educational programs. By January 2010, some 18 million accounts were registered, although there are no reliable figures for actual long term consistent usage [82]. More recently, Second Life has been used in a number of medical and health educational projects. "Nutrition Game" was designed to increase awareness about the health effects of eating fast food. Players choose foods from different restaurants to simulate their daily intake. Information about each item is available including statistics as to how their health would be affected by this style of eating. Initial results show that the game has an effect on a players' conception of fast food and will likely influence their food choices [83].

The Virtual Neurological Education Centre (VNEC) http://healthinfoisland.blogspot.com/2007/03/virtual-neurological-education-centre.html provides a virtual online experience where users expose themselves to common symptoms encountered by a person suffering from neurological symptoms (e.g., motor incoordination, abnormal sensation) [84]. It aims to achieve an interactive and immersive experience such that the simulated sensations make users feel as though they are in another reality. This limits the users' independence, movement and coordination with the aim of helping them empathize with the effects of neurological disabilities. It also aims to give people with neurological disabilities opportunities to socialize in an environment without their normally felt restrictions.

One of the more intriguing uses of Second Life, and highly relevant in the context of empowerment is the creation of a space for the "Autistic Liberation Front" http://slurl.com/secondlife/Porcupine/29/177/113. It was conceived and developed by two women with Asperger's Syndrome as a forum for people with this condition to organize, educate, and advocate for themselves. Another example of its use for empowerment by and for people with disabilities is Live2Give, a Second Life island for cerebral palsy. In keeping with the mandate of the "social action" quadrant, these Second Life applications have been formulated and driven by people with disabilities to bring them together, and to give them the means to help each other cope with common struggles [85]. The experience appears to be empowering and to drastically change how people with disabilities perceive themselves and their role in the world [86].

6 Discussion

TalkingITGlobal: Inspire, Involve, Inform is an example of a United Nations' project that develops and empowers youth by using ICT and advanced technologies in various countries. One of its members, Nabil, a young student with a disability who is enrolled in a computer literacy class, remarked: "I want to show the world that though I am disabled, I am not disqualified". He further stressed the potential of ICT to promote and develop the lives of persons with disabilities, especially where poverty and disability are major contributors to marginalization. Nabil summarized the potential of ICT and other advanced technologies to "open new vistas for disabled people and make it a different world" by promoting and developing the lives of persons with disabilities [87].

In this chapter, we highlighted the digital gap between people with and without disabilities [19–22], and recognized the need to take advantage of ways in which ICT can reduce this gap. In order to establish a framework for ICT driven social change, we applied Churchman and Barak's [38] adaptation of Jeffries' [37] theoretical model of social change in order to identify technology's potential to enhance the empowerment of people with disabilities. We used concrete examples to illustrate the relevance of the model's four patterns of change – social planning, social development, social action and social reform. We examined the possibility of creating social change for people with disabilities by enabling empowered users of advanced technologies. The social change theoretical framework [38], with its two axes and four quadrants (cf. Fig. 2), showed how all stakeholders, including ICT users with disabilities, family members, service providers, ICT developers, researchers and policy makers, can understand and implement the four different ways to achieving the 'different world' described above by Nabil [87]. Locating a particular social program or rehabilitation intervention into one of the four quadrants helps us to identify how technology is currently being used, and to consider whether a different approach (e.g., social reform versus social planning) would perhaps better suit as the targeted persons, groups or community. It also enables us to identify barriers and facilitators to ICT use when the goal is social change, and to examine ways to acquire the internal and external resources needed to shift from one quadrant to another. Furthermore, it can lead to improvements in the outcome measures of the observed or reported social change and empowerment of people with disabilities, facilitated by using ICT.

Research is needed to validate the use of this model to understand the process of empowerment for people with disabilities. Given the variability of the target population and the need for both objective and subjective outcome measures, a mixed research design including both quantitative and qualitative tools seems warranted. The abilities, attitudes and social views of people with disabilities prior to and following the interventions highlighted in the above examples would provide evidence as to the model's validity.

A question arises as to whether the goal of the social change theory is to achieve social change only by reaching the 'social action' quadrant. As illustrated by the examples presented above, it is our contention that each quadrant serves an important developmental and social role, as long as each project is analyzed to determine that it suits a particular quadrant, and that our partners with disabilities are empowered "nothing about us, without us" [26].

Such an approach leads to several recommendations of using the framework for aiming at the desired social change facilitated by ICT usage. First, it is necessary to identify all the stakeholders in the matrix. Second, we must identify where the stakeholders are along the axes of change from the secure well-known stability, and where they are along the empowerment-users axis. Such an in-depth analysis will provide a picture of where a given project is located, in order to set goals where it should head. This will lead to a set of outcome measures that will evaluate the change in the project. Finally, the most important recommendation is that developing the most advanced and suitable technology for people with disabilities must be accompanied with empowerment courses. Those courses should focus on empowering the ICT users with disabilities, but at the same time focus on teaching other stakeholders how to facilitate empowerment. Without a focus on empowerment, it is highly likely that the ICT illiteracy of people with disabilities will remain, and that social change will fail.

References

1. Hurst, R., and Albert, B.: 2006. The social model of disability, human rights and development cooperation. In: Albert, B. (ed.), In or Out of the Mainstream? Lessons from Research on Disability and Development Cooperation, pp. 24–39. Leeds, UK: Disability Press.
2. Longmore, P. K.: 2003. Why I Burned My. Book and Other Essays on Disability, Philadelphia, PA, Temple University Press, 278 pp.
3. Poetz, C. L.: 2003. Reflections on 30 years of involvement in self-advocacy. J. Intell. Dev. Disabil. 28, 84–87.

4. Reiter, S.: 2008. Disability from a humanistic perspective: Towards a better quality of life. New York: Nova Science Publishers.
5. Preiser, W. F. E., and Ostroff, E. (eds.): 2001. Universal Design Handbook, McGraw-Hill, New York, USA.
6. Sehgal, R., and Stewart, G.: 2006. Using Qualitative Analysis for Deriving Evidence Based Construct Definition: A Case Narrative of 'User Empowerment. In: Proceedings of the 3rd International Conference on Qualitative Research in IT & IT in Qualitative Research, 27–29 November 2006, Australia, Queensland, Brisbane.
7. Page, N., and Czuba, C.E.: 1999. Empowerment, What is it? Journal of Extension, [On-line], 37, (5) 1–7, http://www.joe.org/joe/1999october/comm1.html.
8. Kreisberg, S.: Transforming power: Domination, empowerment, and education. Albany, NY: State University of New York Press (1992)
9. Rappoport, J.: 1984. Studies in empowerment: Introduction to the issue. Prev. Hum. Serv. 3, 1–7.
10. Rogers, E. S., Chamberlin, J., Ellison, M. L., and Crean, T.: 1997. A consumer-constructed scale to measure empowerment among users of mental health services. Psychiatr. Serv. 48, 1042–1047.
11. Rosenfield, S., Neese-Todd, S.: 1993. Elements of a psychosocial clubhouse program associated with a satisfying quality of life. Hosp. Community Psych. 44, 76–78.
12. Wowra, S., McCarter, R.: 1999. Validation of the empowerment scale with an outpatient mental health population. Psychiatr. Serv. 50, 959–961.
13. Florin, P., Wandersman, A.: 1190. An introduction to citizen participation, voluntary organizations, and community development: insights for empowerment through research. Am. J. Commun. Psychol. 18(1), 41–54.
14. Wilson, P.: 1996. Empowerment: Community economic development from the inside out. Urban. Stud. 33(4–5), 617–630.
15. Zimmerman, M.A.: 2000. Empowerment theory: psychological, organizational and community levels of analysis. In: Rappaport, J. and Seidman, E. (eds.), Handbook of Community Psychology, pp. 43–63. Plenum Press, New York.
16. Spreitzer, G.M.: 1995. Psychological empowerment in the workplace: dimensions, measurement, and validation. Acad. Manage. J. 38 (5), 1442–1465.
17. Doll, W. J., Deng, X., Metts, G. A.: 2003.User Empowerment in Computer mediated Work. Paper presented at the ISOneWorld 2003, Las Vegas, Nevada, USA.
18. Amichai-Hamburger, Y., MaKenna, K.Y.A., and Tal, S.A.: 2008. E-empowerment: Empowerment by the Internet. Comput. Human. Behav. 24 (5), 1776–1789.
19. G3ict: 2007. The Accessibility Imperative. A Division of World Times, Inc, http://www.g3ict.com
20. Vicente M.R., and López A.J.: 2009. Are people with disabilities losing the ICT Revolution? Some evidence on European countries. The Open Soc. Sci. J. 2, 78–81.
21. Dobransky, K., Hargittai, E.: 2006. The disability divide in Internet access and use. Inform. Commun. Soc. 9, 313–34.
22. Research and Training Center on Disability in Rural Communities Disability and the Digital Divide: 2006. Comparing Surveys with Disability Data, http://rtc.ruralinstitute.umt.edu/TelCom/Divide.htm
23. Blair, J.: 2006. A computer and internet future: enabling inclusion? Learn. Disabil. Pract. Res. 9(9), 31–37.
24. European Commission: 2007. Empirica and Work Research Center. Measuring e-accessibility in Europe. Brussels: European Commission, http://ec.europa.eu/information_society/activities/einclusion/library/studies/meac_study/index_en.htm
25. Zimmerman, M.A., Warschausky, S.: 1998. Empowerment Theory for Rehabilitation Research: Conceptual and Methodological Issues. Rehabil. Psychol. 43 (1), 3–16.
26. Riva, G., and Gaggioli, A.: 2009. Rehabilitation as empowerment: The role of advanced technologies. In: Gaggioli, A., Keshner, E., Weiss, P.L., Riva, G. (eds.) Advanced Technologies in Rehabilitation: Empowering Cognitive, Physical, Social and Communicative Skills through

Virtual Reality, Robots, Wearable Systems and Brain-Computer Interfaces. IOS Press: Studies in Health Technology and Informatics.
27. Charlton, J.I.: 2000. Nothing about us without us: disability oppression and empowerment. Berkley, California: University of California Press.
28. Chamberlin, J.: 1997. A working definition of empowerment. Psychiatr. Rehabil. J. 20(4), 43–46.
29. Bunning, K., Heath, B., Minnion, A.: 2009. Communication and Empowerment: A Place for Rich and Multiple Media? J. Appl. Res. Intellect. 22 (4), 370–379.
30. Lopez-Claros, A., and Zahidi, S.: 2005. Women's empowerment: measuring the global gender gap, The World Economic Forum website, http://www.weforum.org/pdf/Global_Competitiveness_Reports/Reports/gender_gap.pdf
31. Valaitis, R.K.: 2005. Computers and the Internet: Tools for youth empowerment. J. Med. Internet. Res. 7(5), e51, http://www.ncbi.nlm.nih.gov/pmc/articles/PMC1550685/
32. Rizzo, A.A., Buckwalter, J.G., Neumann, U.: 1997. Virtual reality and cognitive rehabilitation: a brief review of the future. J. Head. Trauma. Rehabil. 12, 1–15.
33. Rizzo, A.A., Kim, G.: 2005. A SWOT analysis of the field of Virtual Rehabilitation and Therapy. Presence-Teleop. Virt. 14, 1–28.
34. Weiss, P.L., Kizony, K., Feintuch, U. and Katz, N.:2006. Virtual Reality in Neurorehabilitation. In: Selzer, M. E., Cohen, L., Gage, F.H., Clarke, S., & Duncan, P. W. (eds.) the Textbook of Neural Repair and Neurorehabilitation, vol. 2, pp.182–197. Cambridge University Press.
35. Schultheis, M.T., Rizzo, A.A.: 2001. The application of virtual reality technology for rehabilitation. Rehabil. Psychol. 46, 296–311.
36. Rose, F. D., Brooks, B. M. and Attree, E. A.: 2000. Virtual reality in vocational training of people with learning disabilities. In: Proceedings of the 3rd International Conference on Disability, Virtual Reality and Associated Technologies (ICDVRAT), Sardinia, pp. 129–135.
37. Jack, D., Boian, R., Merians, A., Tremaine, M., Burdea, G.C., Adamovich, S.V., Recce, M., Poizner, H.: 2001.Virtual reality- enhanced stroke rehabilitation. IEEE Trans. Neural. Syst. Rehabil. Eng. 9, 308–318.
38. Jeffries, A.: 1996. Modelling community work: an analytic framework for practice. J. Community Pract. 3 (3/4), 101–125.
39. Churchman, A., and Barak, A.: 2003. Empowerment and partnership– a deceiving terminology In Churchman, A., and Sadan, E. (eds.) Participation: your way to make a difference, pp. 106–124. Tel Aviv: Hakibbutz Hameuchad Publication House Ltd.
40. Weil, M., and Gamble, D.N.: 2005. Evolution, models, and changing context of community practice. In: Weil, M. (ed.) The handbook of community practice. pp.117–150. Sage Publication, Inc.
41. Blomquist, K. B.: 2006. Health, education, work, and independence of young adults with disabilities. Orthop. Nurs. 25(3), 168–187.
42. Hudson, B.: 2003. From adolescence to young adulthood: The partnership challenge for learning disability services in England. Disabil. Soc. 18(3), 259–276.
43. Lotan, M., Isakov, E., Kessel, S., Merrick, J.: 2004. Physical fitness and functional ability of children with intellectual disability: effects of a short-term daily treadmill intervention. The Sci. World. J. 4, 449–457.
44. Pitetti, K. H., Boneh, S.: 1995. Cardiovascular fitness as related to leg strength in adults with mental retardation. Med. Sci. Sports. Exerc. 27(3), 423–428.
45. Halle, J. W., Gabler-Halle, D., Chung, Y. B.: 1999. Effects of a peer mediated aerobic conditioning program on fitness levels of youth with mental retardation: two systematic replications. Ment. Retard. 37(6), 435–448.
46. Fernhall, B., Tymeson, G. T.: 1998. Validation of a cardiovascular fitness field test for adults with mental retardation. Adapt. Phys. Activ. Q. 5, 49–59.
47. Dattilo, J. Schleien, S.J.: 1994. Understanding leisure services for individuals with mental retardation. Ment Retard. 32, 53–9.

48. Thomas, A.D., Rosenberg, A.: 2003. Promoting community recreation and leisure. Pediatr. Phys. Ther. 15, 232–246.
49. Van Naarden Braun, K., Yeargin-Allsopp, M., and Lollar, D.: 2006. Factors associated with leisure activity among young adults with developmental disabilities. Res. Dev. Disabil. 27, 567–583.
50. Frey, G. C., Buchanan, A. M., Rosser Sandt, D. D.: 2005. "I'd Rather Watch TV": An Examination of Physical Activity in Adults with Mental Retardation. Ment. Retard. 43, 241–254.
51. Hayden, M.F., Lakin, K.C., Hill, B.K., Bruininks, R.H.: 1992. Social and leisure integration of people with mental retardation in foster homes and small group homes. Educ. Train. Ment. Retard. 27, 187–99.
52. Langone, J.: 1996. Mild mental retardation. In: McLaaughlin, P.J., Wehman, P. (eds.) Mental retardation and developmental disabilities. pp. 113–129. Austin, Texas: Pro-ed.
53. Schleien, S.J., Ray, M.T.: 1998. Community recreation and persons with disabilities. Baltimore: Paul H. Brookes.
54. Specht, J., King, G., Brown, E., Foris, C.: 2002. The importance of leisure in the lives of persons with congenital physical disabilities. Am. J. Occup. Ther. 56, 436–445.
55. Brett, A.: 1997. Assistive and adaptive technology: Supporting competence and independence in young children with disabilities. Dimensions of Early Childhood, 25, 14–20.
56. Bryen, D.N., Slesaransky, G., and Baker, D.B.: 1995. Augmentative communication and empowerment supports. Augment. Altern. Commun. 11, 79–88.
57. Mirenda, P., Mathy-Laikko, P.: 1989. Augmentative and Alternative Communication applications for persons with severe congenital communication disorders: An introduction. Augment. Altern. Commun. 1, 3–13.
58. Fine, A.H. Fine, N.M.: 1996. Therapeutic recreation for exceptional children: Let me in I want to play (2nd ed.). Springfield, IL: Charles C Thomas.
59. Raz, S.: 1998. Establish a multi-purpose model for the rehabilitation of children with special needs. World Leisure and Recreation, 1, 30–34.
60. Suto, M.: 1998. Leisure in occupational therapy. Can. J. Occup. Ther. 65, 271–278.
61. Standen, P.J., Brown, D.J.: 2005. Virtual reality in the rehabilitation of people with intellectual disabilities: review. CyberPsychol. Behav. 8, 272–282.
62. Weiss, P.L., Sveistrup, H., Rand, D. and Kizony, R.: 2009. Video capture virtual reality: A decade of rehabilitation assessment and intervention. Phys. Ther. Rev. 14, 307–321.
63. Lotan, M., Yalon-Chamovitz, S., Weiss P.L.: 2009. Improving physical fitness of individuals with intellectual and developmental disability through a Virtual Reality Intervention Program. Res. Dev. Disabil. 30, 229–239.
64. Lotan, M., Yalon-Chamovitz, S. and Weiss, P.L. Virtual Reality as means to improve physical fitness of individuals at a severe level of intellectual and developmental disability. Res. Dev. Disabil. In press.
65. Ruuskanen, J. M., Parkatti, T.: 1994. Physical activity and related factors among nursing home residents. J. Am. Geriatr. Soc. 42(9), 987–991.
66. Temple, V. A.: 2007. Barriers, enjoyment, and preference for physical activity among adults with intellectual disability. Int. J. Rehabil. Res, 30(4), 281–287.
67. Fox, R. A., Rosenberg, R., and Rotatori, A.: 1985. F. Parent involvement in a treatment program for obese retarded adults. Behav Ther Exp Psychiatry, 16(1), 45–48.
68. Chamberlin, J.: 1997. A working definition of empowerment. Psychiatr. Rehabil. J. 20 (4), 43–46.
69. Brennan, D., Georgeadis, A., Baron, C.: 2002. Telerehabilitation tools for the provision of remote speech-language treatment. Top. Stroke Rehabil. 8(4), 71–78.
70. Samoocha, D., Bruinvels, D.J., Anemal, J.R., Steenbeek, R., van der Beek, A.K.: 2009. Empowerment of disability benefit claimants through an interactive website: design of a randomized controlled trial. BMC Med. Inform. Decis. Mak. 9, 23.
71. Brennan, D.M., Barker, L.M.: 2008. Human factors in the development and implementation of telerehabilitation systems. J. Telemed Telecare. 14, 55–58.

72. Jennett, P.A., Affleck Hall, L., Hailey, D., Ohinmaa, A., Anderson, C., Thomas, R., Young, B., Lorenzetti, D. Scott, R.E.: 2003. The socio-economic impact of telehealth: a systematic review. J. Telemed. Telecare. 9, 311–320.
73. Coulter, A., and Ellins, J.: 2006. Patient-focused interventions: A review of the evidence. The Health Foundation , London http://www.health.org.uk/publications/patient-focused-interventions/ Retrieved 2011-05-22.
74. Mountain, G.A., Ware, P.M., Hammerton, J., Mawson, S.J., Zheng, H., Davies, R., Black, N.D., Zhou, H., Hu, H., Harris, N., and Eccleston, C.: 2006. The SMART Project: A user led approach to developing and testing technological applications for domiciliary stroke rehabilitation. In: Clarkson, P., Langdon, J., and Robinson, P. (eds.) Designing Accessible Technology, Springer-Verlag London.
75. Burgstahler, S., and Cronheim, D.: 2001. Supporting peer-peer and mentor-protege relationships on the internet. J. Res. Technol. Educ. 34(1), 59–74.
76. Seeger, K.L.: 2007. Mentoring youth with disabilities: The mentor's lived experiences. Unpublished master's thesis, The School of Human Resource Education and Workforce Development, Louisiana State University, U.S.A.
77. Shpigelman, C., Reiter, S. and Weiss P.L.: 2009. A conceptual framework for electronic socio-emotional support for people with special needs. Int. J. Rehabil. Res. 32:301–308.
78. Andersson, G., Bergström, J., Holländare, F., Carlbring, P., Kaldo, V., and Ekselius, L.: 2005. Internet-based self-help for depression: randomized controlled trial. Br. J. Psychiatry. 187, 456–461.
79. Barak, A., and Dolev-Cohen, M.: 2006. Does activity level in online support groups for distressed adolescents determine emotional relief? Couns. Psychotherapy Res. 6(3), 186–190.
80. van Uden-Kraan, C.F., Drossaert, C.H.C., Taal, E., Shaw, B.R., Seydel, E.R., van de Laar, M.A.F.J.: 2008. Empowering processes and outcomes of participation in online support groups for patients with breast cancer, arthritis, or fibromyalgia. Qual. Health Res. 18(3), 405–417.
81. Bowker, N.I., and Tuffin, K.: 2007. Understanding positive subjectives made possible online for disabled people. NZ. J. Psychol. 36(2), 63–71.
82. Current user metrics for Second Life". http://secondlife.com/xmlhttp/secondlife.php. Retrieved 2010-02-19
83. Cooper, T.: 2007. Nutrition Game. Second Life Education Workshop 2007. 46–50.
84. Kamel Boulos, M.N., Hetherington, L., Wheeler, S.: 2007. Second Life: An overview of the potential of 3-D virtual worlds in medical and health education. Health Info. Libr. J. 24, 233–245.
85. Gorini, A., Gaggiolil, A., Vigna, C., Riva, G.: 2008. A Second Life for eHealth: Prospects for the use of 3-D virtual worlds in clinical psychology. J. Med. Internet Res. 10(3), e21.
86. Lester, J.: All about Live2Give. Live2Give: An innovative Online Community for people dealing with Cerebral Palsy and similar conditions. Braintalk Blog. http://braintalk.blogs.com/live2give/2005/01/all_about_live2.html
87. eLeaders for Youth and ICT (2009). A digital shift: Youth and ICT for development. Global for ICT development. United Nations, Department of Economic and Social Affairs, Global Alliance for ICT and Development, New York. http://unpan1.un.org/intradoc/groups/public/documents/gaid/unpan036084.pdf

FSC
www.fsc.org
MIX
Papier | Fördert
gute Waldnutzung
FSC® C083411

Kolibri 360 GmbH
Ferdinand-Jühlke-Straße 7
99095 Erfurt, Deutschland
produktsicherheit@kolibri360.de